Iosias Jody (Ed.)

Nokia X2-01

Iosias Jody (Ed.)

Nokia X2-01

Feature phone, Series 40, Mobile operating system

Cred Press

Imprint

Publisher:
Cred Press is a trademark of
International Book Market Service Ltd., 17 Rue Meldrum, Beau Bassin, 1713-01 Mauritius
Email: info@bookmarketservice.com
Website: www.bookmarketservice.com

Published in 2012

Printed in: U.S.A., U.K., Germany. This book was not produced in Mauritius.

ISBN: 978-620-1-90298-5

Contents

Articles

References

Nokia_X2-01

Manufacturer	Nokia
Series	Nokia Xseries
Form factor	Candybar
Dimensions	119.4 x 59.8 x 14.3 mm
Weight	107.5 g (with battery)
Operating system	Series 40 V 08.71 FOTA (firmware update over the air)
Memory	Internal - 55 MB, 64 MB RAM, 128 MB ROM
Storage	up to 2000 entries
Removable storage	Expandable to 8GB
Battery	Standard battery, Li-Ion 1020 (BL-5C) Stand-by Up to 480 h Talk time Up to 4 h 30 min
Data inputs	FULL QWERTY,Dedicated music and messaging keys
Display	2.4" Resolution: 320 x 240 pixels (TFT) 262,000 colors
Rear camera	VGA
Ringtones & notifications	AAC, eAAC+, MP3, WMA, MIDI
Connectivity	Bluetooth version 2.1 with Enhanced Data Rate,High-Speed USB 2.0
Development status	2011-Present

The **Nokia X2-01** is a low-cost feature phone with an Nokia S40 mobile operating system released under the X-series line of phones by Nokia. It features a full QWERTY keyboard. It is being advertised as an entry-level messaging & Music phone.

Features

The Nokia X2-01 is installed with Ovi Mail & Ovi Chat where users can set up email and chat accounts straight from the device. The X2-01 also has a VGA camera, 2.4 inch screen, and support for up to 16GB of storage on a MicroSD card. The Phone is available in various colours.

Specifications

- General
  - 2G Network: GSM 850 / 900 / 1800 / 1900
  - Announced: 2010, November
  - Status: Available. Released 2011, January
- Body

    - Dimensions: 119.4 x 59.8 x 14.3 mm, 86.6 cc
    - Weight: 107.5 g
    - Keyboard: QWERTY
- Display
    - Type: TFT, 256K colors
    - Size: 320 x 240 pixels, 2.4 inches (~167 ppi pixel density)

-Dedicated music key

- Sound
    - Alert types: Vibration, Polyphonic(64), WAV, MP3 ringtones
    - Loudspeaker: Yes
    - 3.5mm jack: Yes
- Features
    - Primary: VGA, 640x480 pixels
    - Video: Yes, QVGA@24fps
    - Secondary: No
- Features
    - Messaging: SMS, MMS, Email
    - Browser: WAP 2.0/xHTML, HTML (Opera Mini)
    - Radio: Stereo FM radio with RDS
    - Games: Yes + downloadable
    - GPS: No
    - Java: Yes, MIDP 2.1
    - Colors: Red, Deep Grey, Silver, Lilac, And Azure

- MP4/H.264/H.263/WMV player - MP3/WAV/WMA/AAC player - Organizer - Voice memo - Predictive text input

- Battery
    - Battery: Standard battery, Li-Ion 1020 (BL-5C)
    - Stand-by: Up to 480 h
    - Talk time: Up to 4 h 30 min

Feature_phone

A **feature phone** is a mobile phone which at the time of manufacture is not considered to be a smartphone, but nevertheless has additional functions over and above standard mobile services. It is intended for customers who want a lower-price phone without all the features of a smartphone.

In 2011, feature phones accounted for 60 percent of the mobile telephones in the United States[1] and 70 percent of mobile phones sold worldwide.[2]

A feature phone

Difference between smartphone and feature phone

Although feature phone is the term used to describe low-end devices and smartphone is used to describe high-end devices, there is no official definition to distinguish the two.[3][4] Smartphone and feature phone are not mutually exclusive categories.[5]

The price difference between a smartphone and feature phone is often used to distinguish the two devices. As of March 2012, Canadian cellular service providers offer the choice of purchasing smartphone upfront for $450-650 CAD on "no term" (month-by-month), or signing 3-year voice and data contract to waive most of the handset purchase cost by (there are no waivers for a voice-only plan). The no term price for a feature phone, by contrast, is typically half or even less than that of a smartphone, and this cost can be waived with a 3-year voice-only plan.[6][7][8]

A complication in distinguishing between smartphones and feature phones is that over time the capabilities of new models of feature phones can increase to exceed those of phones that had been promoted as smartphones in the past. Because technology changes rapidly, what was a smartphone ten years ago may be considered only a feature phone today. For example, today's feature phones typically also serve as a personal digital assistant (PDA) and portable media player and have capabilities such as camera phones, touchscreen, GPS navigation, Wi-Fi, and mobile broadband access.

Another significant difference between smartphones and feature phones is that the advanced application programming interfaces (APIs) on smartphones for running third-party applications[9] can allow those applications to have better integration with the phone's OS and hardware than is typical with feature phones. In comparison, feature phones more commonly run on proprietary firmware, with third-party software support through platforms such as Java ME or BREW.[10] While advanced APIs appear on smartphones first, they are gradually moving to feature phones.[11]

References

[1] Don Kellogg (1 September 2011). "40 Percent of U.S. Mobile Users Own Smartphones; 40 Percent are Android" (http://blog.nielsen.com/nielsenwire/online_mobile/40-percent-of-u-s-mobile-users-own-smartphones-40-percent-are-android/). *Nielsen Company*. .

[2] (http://www.google_ca/url?sa=t&rct=j&q=feature phone&source=web&cd=2&ved=0CFwQFjAB&url=http://www.zdnet.com/blog/cell-phones/nokias-continued-feature-phone-focus-may-be-one-of-their-smartest-moves/7215&ei=JldbT5b_Fcar0AG_wJzeBQ&usg=AFQjCNEEHS4L6VDciCJm5SGx67GAIZTH7g&cad=rja)

[3] "Feature Phone" (http://www.phonescoop.com/glossary/term.php?gid=310). *Phone Scoop*. . Retrieved 9 May 2010.

[4] Andrew Nusca (20 August 2009). "Smartphone vs. feature phone arms race heats up; which did you buy?" (http://www.zdnet.com/blog/gadgetreviews/smartphone-vs-feature-phone-arms-race-heats-up-which-did-you-buy/6836). ZDNet. .

[5] http://www.brighthand.com/article/Study_Says_Smartphones_Will_Outsell_Handhelds/

[6] (http://www.electronista.com/articles/07/12/13/85k.canada.cell.data.bill/)

[7] (http://www.bell.ca/Mobility/Smartphones_and_mobile_internet_devices)

[8] [www.rogers.com/web/link/wirelessBuyFlow?forwardTo=PhoneThenPlan&productType=normal]

[9] "Smartphone definition from PC Magazine Encyclopedia" (http://www.pcmag.com/encyclopedia_term/0,2542,t=Smartphone&i=51537,00.asp). *PC Magazine*. . Retrieved 2011-12-15.

[10] "Smartphone" (http://www.phonescoop.com/glossary/term.php?gid=131). *Phone Scoop*. . Retrieved 2011-12-15.

[11] (http://www.pocketgamer.biz/r/PG.Biz/Connect2Media+news/feature.asp?c=38677)

Series_40

Series 40 is a software platform and application user interface (UI) software on Nokia's broad range of mid-tier feature phones, as well as on some of the Vertu line of luxury phones. It is the world's most widely used mobile phone platform and found in hundreds of millions of devices.[1] Nokia announced on 25 January 2012 that the company has sold over 1.5 billion S40 devices. [2] S40 has more features than the Series 30, which is a very basic OS. Neither of them are based on Symbian.

Series 40-based Nokia 6300

History

Series 40 was officially introduced in 1999 with the release of the Nokia 7110. It had a 96 × 65 pixel monochrome display and was the first phone to come with a WAP browser. Over the years, the S40 UI has evolved from a low resolution UI to a high resolution color UI with an enhanced graphical look. The third generation of Series 40 that became available in 2005 introduced support for devices with resolutions as high as QVGA (240×320).[3] It is possible to customize the look-and-feel of the UI via comprehensive themes.[4] A list of all Series 40 devices can be found on the Nokia web site.[5]

In 2012, the new Nokia Asha mobile phones 200/201, 302, 303 and 311 were released and all use Series 40.[5]

Features

Home screen of Nokia S40 v10.80 running on Nokia 6303i

Applications

It provides communication applications such as telephone, internet telephony (VoIP), messaging, email client with POP3 and IMAP4 capabilities and Web browser; media applications such as camera, video recorder, music/video player and FM radio; and phonebook and other personal information management (PIM) applications such as calendar and tasks. Basic file management, like in Series 60, is provided in the Applications and Gallery folders and subfolders. Gallery is also the default location for files transferred over Bluetooth to be placed. User-installed applications on Series 40 are generally mobile Java applications. Flash Lite applications are also supported, but mostly used for screensavers.[6]

Web browser

The integrated web browser can access most web content through the service provider's XHTML/HTML gateway. The latest version of Series 40, called Series 40 6th Edition, introduced a new browser based on the WebKit open source components WebCore and JavaScriptCore. The new browser delivers support for HTML 4.01, CSS2, JavaScript 1.5, and Ajax. Also, like the higher-end Series 60, Series 40 can run the Opera Mini web browser to enhance the user's web browsing experience.

Synchronization

Support for SyncML synchronization with external services of the address book, calendar and notes is present. However with many S40 phones, these synchronization settings must be sent via an OTA text message.

Technical

Software platform

Series 40 is an embedded software platform that is open for software development via standard or de-facto content and application development technologies. It supports Java MIDlets, i.e. Java MIDP and CLDC technology, which provide location, communication, messaging, media, and graphics capabilities.[7] S40 also supports Flash Lite applications.[6]

Operating system

Series 40 is a simpler operating system than the higher end S60. Because S40 devices do not support true multi-tasking and do not have a native code API for third parties, its user interface may appear to be more responsive and faster than the other Nokia platforms Symbian S60 on similar hardware.[8]

References

[1] "Forum Nokia - Nokia Series 40 Platform" (http://www.forum.nokia.com/Devices/Series_40/). Nokia. . Retrieved 2010-10-27.

[2] "Nokia has sold over 1.5 billion Series 40 phones" (http://www.esphoneblog.com/2012/01/25/nokia-has-sold-over-1-5-billion-series-40-phones/). . Retrieved 2012-01-25.

[3] "Series 40 UI Style Guide – Forum Nokia" (http://www.forum.nokia.com/info/sw.nokia.com/id/73e935fe-8b59-43b2-ab3e-1c5f763672db/Series_40_UI_Style_Guide.html). Nokia. . Retrieved 2008-09-26.

[4] "Carbide.ui Theme Edition (can be used to create S40 themes) – Forum Nokia" (http://www.forum.nokia.com/Library/Tools_and_downloads/Other/Carbide.ui/). Nokia. . Retrieved 2011-05-16.

[5] "Device specifications, filtered for Series 40" (http://www.developer.nokia.com/Devices/Device_specifications/?filter1=s40). Nokia. . Retrieved 2012-06-01.

[6] "Working with Nokia Series 40 Flash Lite content – Adobe Developer Center" (http://web.archive.org/web/20080518220200/http://www.adobe.com/devnet/devices/articles/nokia_series40_pt1_print.html). Adobe Systems. Archived from the original (http://www.adobe.com/devnet/devices/articles/nokia_series40_pt1_print.html) on 2008-05-18. . Retrieved 2008-09-26.

[7] "Developing Scalable Series 40 Applications, A Guide for Java Developers" (http://www.pearsoned.co.uk/BOOKSHOP/detail.asp?item=100000000075919). Addison-Wesley. . Retrieved 2008-09-26.

[8] "Comparing Series 40 against S60 (as of 2007) – All about Symbian" (http://www.allaboutsymbian.com/features/item/Series_40_vs_S60.php). . Retrieved 2008-09-26.

Nokia_phone_series

Nokia's nomenclature can be traced back in 2005, when the Nseries line was introduced.[1] Because of the demands and peak of that line, Nokia again introduced another series of phones named Eseries,[2] made mostly for the enterprise market and to compete with Palm and BlackBerry.[3] In 2009, Nokia dropped their four-number naming scheme[4] and focused on their two existing lines, with additional two introduced during Nokia World[5] being the Xseries which is for entertainment and music enthusiasts,[6] while Cseries is what Nokia considers their "core" range of products.[7] A Chinese-exclusive range of devices named Tseries was introduced on 13 June 2011.[8] Two more series were introduced at Nokia World 2011 *Asha* which means "hope" and *Lumia* which means "light."[9][10] Other series of phones include the N-Gage,[11] XpressMusic,[12] Navigator series, Prism,[13] Supernova[14] and the fashion-oriented Distinctly Bold[15] and L'Amour Collection.[16]

Notable Nokia phones

Nokia Nseries

Nokia Nseries

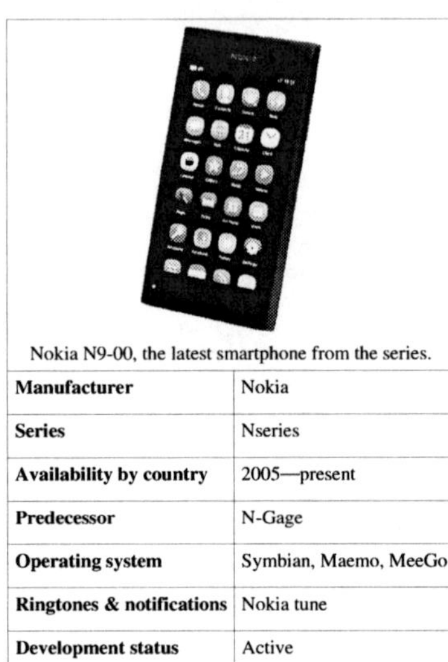

Nokia N9-00, the latest smartphone from the series.

Manufacturer	Nokia
Series	Nseries
Availability by country	2005—present
Predecessor	N-Gage
Operating system	Symbian, Maemo, MeeGo
Ringtones & notifications	Nokia tune
Development status	Active

Nokia Nseries is a product family composed exclusively of smartphones and Internet tablets. Mobile devices included in the series include digital multimedia services and other advanced features. The company's flagship products come from this series, and this is where most of Nokia Research Center's technologies and innovations are applied.[17]

- **Nokia N70**[18] is one of the first products successfully launched from the series. It has introduced a sliding camera cover on its rear view, protecting the lens and triggers the camera feature when opened, which was later used in other phones created by Nokia.[19] High resolution 2-megapixel camera with video capture, Bluetooth, FM radio, RealPlayer media viewing and support for WAP 2.0 are also the notable features offered by this device.[20]
- **Nokia N71**[21] is a phone in clamshell form factor,[22] mostly recommended for consumers who search for enhanced ergonomics due to its button arrangement.[23] It supports SMS, MMS, voice and video calling, Internet browsing, Bluetooth wireless technology, among many other features. Like the Nokia N70, it has a 2-megapixel camera.[24] To connect with PC, it uses the proprietary Pop-Port medium.
- **Nokia N72**[25] is a stylish Nseries device that comes in two designs: gloss black[26] and pearl pink.[27] It only runs through the GSM network, and has no front-facing camera.[28] The device takes from Nokia N70's design cues.
- **Nokia N73**[29] is one of the range's best-selling smartphones.[30] Apart from the more compact design, it also has improved its camera compared to the Nokia N72, packing 3.2-megapixels with Carl Zeiss lens.[31] It is capable of video calling, and was the first device to be termed the "multimedia computer."[32] The camera is praised for taking sharp, clear and crisp photos.[33]

List of Eseries devices

Nokia Eseries

Manufacturer	Nokia
Series	Eseries
Availability by country	2005—present
Predecessor	Communicator
Operating system	Symbian, Series 30
Ringtones & notifications	Nokia tune
Development status	Active

The **Nokia Eseries** (stands for *Executive*) consists of business-oriented smartphones, with emphasis on enhanced connectivity and support for corporate e-mail push services. These phones are for the enterprise market.

The list may also include upcoming devices or others that are previously intended to be a part of the series, but was scrapped or renamed due to Nokia's number-only naming change. These are the phones included in the series:

- **Nokia E50**[34] is a phone intended primarily for the corporate business market.[35] It includes sophisticated e-mail support for Intellisync, BlackBerry Connect, Visto Mobile and ActiveSync Mail for Exchange.[36]
- **Nokia E51**[37] is one of the leaner enterprise devices from the Finnish phone-maker. One of the main highlights of the phone is the multiple one-touch dedicated buttons for email, contacts and calendar applications. These keys are also customizable, recognizing short and long presses.[38]
- **Nokia E52**[39] is an Eseries device with a longer battery life, for consumers looking for an enterprise phone that is well-built and does not drain quickly. The phone can also serve as a data modem through the use of Bluetooth wireless technology.
- **Nokia E55** is a mobile phone for business use, intended to be compact, yet suitable for writing e-mails and other longer texts. This is done via a so-called compact QWERTY keyboard[39] with two characters per key. Designwise, the Nokia E55 is almost exactly identical to its twin, the Nokia E52, and both phones feature exactly the same specifications. The only difference lies in their keypad layout.
- **Nokia E60**[40] is a traditional candybar style smartphone from the Eseries business phone range, a S60 3rd Edition device. The phone had SIP (VoIP) and Wi-Fi support and a unique high resolution display. E60 distinctly had no camera. This is one of the first Eseries devices to be announced.
- **Nokia E61**[41] is a QWERTY business device initially for the European market, but then later released worldwide. It follows the form factor of RIM's BlackBerry devices and offers many work-related features.
- **Nokia E61i**[42] is a follow-up to the Nokia E61 featuring a 2-megapixel camera and D-pad instead of a joystick.[43]
- **Nokia E62**[44] boasts a vibrant and sharp screen, a full QWERTY keyboard, and a full array of wireless options (Bluetooth, UMTS). The Symbian smartphone also has solid call quality, a speakerphone, the ability to view and edit Microsoft Office documents, and robust e-mail capabilities.[45][46]
- **Nokia E63**[47] offers a more wallet-friendly price tag compared with other unlocked smartphones and does so without sacrificing too many features.[48] It offers a full QWERTY keyboard[49] and full e-mail capabilities.[50] The Symbian smartphone has integrated Wi-Fi, Bluetooth, and 3G support.[51]
- **Nokia E65**[52] has a beautiful and compact slider design, and features a 2-megapixel camera, e-mail, and productivity apps. It also offers good call quality and solid talk time battery life, and has Bluetooth and Wi-Fi.[53]
- **Nokia E66**[54] includes an accelerometer which brings animations and transition effects. It is a complete redesign of the Nokia E65.

- **Nokia E70**[55] is a candybar/fold keyboard type smartphone from the Eseries range, announced in October 2005 and released in May 2006 at a cost of approximately $500[56] with prices in July 2007 around $345.
- **Nokia E71**[57] is one of the best-selling Eseries devices to date. It features optimized mobile email and messaging experience with full QWERTY keyboard and pocket-size for one-handed typing, two customizable homescreen views with active stand-by plug-ins and application shortcuts – fast and easy switching between Business and Personal modes, quick access to applications with one-touch keys (Home, Calendar, Contacts, Messaging), intelligent input with auto-completion, auto-correction and learning capabilities ensuring fast and error-free typing and a lot more. The phone won several accolades from respected technology blogs and events including CNET,[58] Mobile Choice Consumer Awards,[59] Wired Magazine[60] and GSMA Awards 2009.[61]
- **Nokia E71x**[62] is the thinnest smartphone available on AT&T before the arrival of Apple's iPhone 4.[63] The phone is affordably priced and offers a long list of features, including 3G, Wi-Fi, Bluetooth, GPS, and a 3.2-megapixel camera, as well as support for AT&T's various services. The QWERTY smartphone is very sleek with high-quality construction, clad in "metal black" color.[64]
- **Nokia E72**[65] with free GPS navigation is a high-performance device tailor-made for seamless business and personal communication.[66] It takes cues from its predecessor, the Nokia E71 and has an improved look of the user interface.[67]
- **Nokia E72i** is a China-only handset made specifically to cater services offered in that region. It also replaced WLAN with WAPI, as it is the standard wireless network available in that country. Most of the features found on the original Nokia E72 were retained.[68]
- **Nokia E73**[69] is a T-Mobile exclusive offering from Nokia. The pre-installed Ovi Maps provides free turn-by-turn navigation.[70] It combines the features of Nokia E71 and Nokia E72, to appeal to the North American market.[71]
- **Nokia E75**[72] ships with Nokia's new messaging client, which features push delivery for all accounts. Other highlights include 3G, Wi-Fi, Bluetooth, and a 3.2-megapixel camera. On front, you will find a 2.4-inch, 16 million-color QVGA display. It's a good size screen considering the smaller chassis, and it's quite sharp and vibrant. In addition, it features a light sensing technology that adjusts the display's brightness depending on what environment you are in. As always, you can customize the homescreen with various themes, wallpaper, and change the font size. Also, like the Nokia E71, you can switch between two different home views depending on whether you are at work or play. In Business mode, the phone gives you immediate access to your e-mail and productivity apps, while switching to the Personal view will bring recreational features, like the media player, N-Gage games, to the forefront.[73]
- **Nokia E90**[74] is equipped with Bluetooth, Wi-Fi, and GPS, and has a solid set of productivity and communication features for the business user. The Symbian smartphone also has a full QWERTY keyboard and 3.2-megapixel camera.[75] The Nokia E90 does not support U.S. 3G networks, and speakerphone volume is a bit weak. The handset is also heavy and bulky.[76]
- **Nokia E5**[77] features IM (instant messenger) that allows access to several chat services or communities simultaneously.[78]
- **Nokia E6**[79] is one of the first two smartphones to be released with the Symbian Anna software update. It includes 50+ enhancements to the operating system, and a new set of icons.[80] It is a compact QWERTY and touch device for business and general users who must get things done while they are on the move.
- **Nokia E7**[81] boasts Microsoft Exchange ActiveSync support alongside a luscious 4-inch touchscreen display, which uses Nokia's ClearBlack technology for improved outdoor visibility. The Nokia E7 also comes with a slide out four-row QWERTY keyboard making it the ideal mobile business device. Despite the addition of a full keyboard, the E7 is only a smidge deeper than the Nokia N8 at 13.6mm thin. The device also boasts an 8-megapixel snapper and comes complete with 16 GB of on-board storage. As a business device, the Nokia E7 comes sporting a raft of security features including business grade device lock and wipe functionality, secure intranet access and remote device management. These come alongside its connectivity and productivity prowess,

with support for Mail for Exchange, Microsoft Communicator Mobile and Microsoft Share Point Server.[82]

- **Nokia E101** is a basic phone with a refreshed version of Series 30, offering a grid-based system of menu icons and capability of storing two SIM cards at a time while remembering the settings made. It offers an enhanced music player, and the ability to store memory cards.[83]

List of Xseries devices

Nokia Xseries

Manufacturer	Nokia
Series	Xseries
Availability by country	2009—present
Predecessor	XpressMusic
Operating system	Symbian, Series 30, Series 40
Ringtones & notifications	Nokia tune
Development status	Active

Nokia Xseries (stands for *Xpress*) is more for younger users. Most of the phones included in the line are feature phones and smartphones that include features focusing social entertainment and music with special dedicated keys, inbuilt storage and other facilities.[84] It succeeded the Nokia XpressMusic brand of phones.

The list may also include upcoming devices or others that are previously intended to be a part of the series, but was scrapped or renamed due to Nokia's number-only naming change. These are the phones included in the series:

- **Nokia X1** is a basic cellular phone with focus on music and entertainment, running on Nokia's proprietary Series 30 platform. The phone offers up to five separate phonebooks, a flashlight and one of the first S30 devices to have a dedicated music player.[85]
- **Nokia X1-01** is a dual-SIM offering from Nokia that aims to connect "the next billion."[86] The device features a dual-standby interface, that allows the user to use both SIMs on the phone at the same time. The phone also has dedicated music player, and a speaker on the back which can deliver up to 106 phon.
- **Nokia X2** packs dual speakers, dedicated music keys, FM stereo and support for up to 16 GB of storage via microSD card make for a nifty mobile music device. Bluetooth 2.1 is supported for wireless music enjoyment whilst hardcore music fans can opt for the 3.5mm headphone jack. Full speed USB 2.0 makes music transfer quick and easy and music can be managed via the Ovi Player PC client or Windows Media Player 11.[87]
- **Nokia X2-01** offers one-click access to music, the media player, FM radio and the Ovi Music service, for unlimited access to a vast library of tunes. It is a cheap QWERTY featurephone running on Series 40 operating system.[88]
- **Nokia X2-03** is the Chinese release of the Nokia X2-01, featuring improvements in its keypad and support for Chinese characters like Pinyin and Mandarin.[89]
- **Nokia X2-05** comes with a 106-phon speaker, enabling superb audio volume and quality. It can also play MP3 and MP4 files, FM radio with recording capabilities and a VGA camera.
- **Nokia X3**[90] is the first Series 40 device to be compatible with the Ovi Store. It is a phone in a sliding form factor that offers great music quality, a 3.2-megapixel camera, a diminutive frame and 2.2-inch screen.
- **Nokia X3-01** is a Chinese specific release, which supports the country's CDMA/EV-DO wireless technology. The phone has a 2.4-inch QVGA-display, FM radio, 3.5mm headset jack, 3.2 megapixel camera with LED flash and support for a number of specific services in China.[91]
- **Nokia X3-02**[92] offers both a touch screen and a traditional 12-button keypad. It is also the world's first mobile phone to run the Series 40 6th Edition operating system, optimized for slate devices. It is one of Nokia's thinnest

devices: it's 9.6mm deep, 106mm tall, 48mm wide and weighs just 78g. The brushed aluminium finish and choice of five colors are also likely to appeal to fashion-conscious mobilists.

- **Nokia X5**[93] is a smartphone running Symbian S60, released in China only, under China Mobile.[94] It makes use of the TD-SCDMA network.
- **Nokia X5-01**[95] features excellent sound quality, dedicated music keys and loud speakers, and will include access to Ovi Music, and to Comes with Music in some regions. There is a full QWERTY keyboard under the dashboard, ready to deliver Tweets, status updates and text messages at your convenience. The device supports all the major IM and webmail services out of the box and also offers easy access to major social networking sites (Facebook, Twitter, Myspace and Hi5). The camera offers 5 megapixels, 4x digital zoom and an LED flash.[96]
- **Nokia X6**[97] features about 16 days standby time, a 5-megapixel camera and Carl Zeiss optics, dual LED flash, TV-out, video editing, online sharing, Nokia Music Store, full Web browser, A-GPS, Ovi Maps, Playlist DJ and Flash Lite support.[98]
- **Nokia X7**[99] offers a unique and durable design that will stand out from the crowd, due to it being made from toughened glass and seamless stainless steel. The 4-inch, capacitive AMOLED display offers 16 million colors and a resolution of 640 x 360 pixels. It is also one of the first phones to have Symbian Anna.[100]
- **Nokia X500** is a Symbian smartphone carrying a 1 GHz processor and the Anna version of the operating system. The phone has colorful exchangeable covers and a 5-megapixel camera.[101] Split-screen messaging, Internet radio, support for a variety of languages and better battery performance were also added to the device.[102]
- **Nokia X600** is the first smartphone from Nokia to feature a 106 phon, fully functional speaker. Running the Symbian Belle operating system, the phone has FM radio transmitter that can forward the music from your phone to your radio receiver, an internal FM antenna that allows the user to listen to FM radio without plugging in the headphones, dedicated music and camera keys and NFC capabilities. The phone features an iconic curved back design available in colors green, pink, white and black, mostly seen from their earlier music devices; it allows better ergonomics and one-hand typing. The Nokia X600 comes with a pentaband radio, and supports 3G data rates up to HSUPA standards, as well as Wi-Fi and Bluetooth.[103] After careful consideration, the phone was cancelled a day before its release, possibly because of the production of the Nokia Windows Phones.[104]

List of Cseries devices

Nokia Cseries

Manufacturer	Nokia
Series	Cseries
Availability by country	2010—present
Operating system	Symbian, Series 30, Series 40
Ringtones & notifications	Nokia tune
Development status	Active

Nokia Cseries (stands for *Core*) is what Nokia calls their "core" range of products, offering different phones from low-end to high-end for various markets.[105]

The list may also include upcoming devices or others that are previously intended to be a part of the series, but was scrapped or renamed due to Nokia's number-only naming change. These are the phones included in the series:

- **Nokia C1** offers flexibility of having two SIM cards in one mobile phone and switch between them at the touch of a button to stay in control of your calls. It also includes a powerful, long-lasting battery, FM radio and torch.[106]

- **Nokia C1-01** enables you to stay in touch with friends, family and the world around you with fast access to your e-mail, instant messaging services, the Internet and Nokia's Ovi Life Tools.[107]
- **Nokia C1-02** lets you stay entertained on the go with the music player – and store more of your favorite tracks with support for up to 32 GB of expandable memory.[108]
- **Nokia C1-03** is a Chinese-specific release combining all the features of Nokia C1-01 with deep integration of traditional Chinese characters and Pinyin.[109]
- **Nokia C2**[110] is a stylish dual SIM phone with Easy Swap technology and great multimedia and messaging features. It also offers a wide range of information covering healthcare, agriculture, education and entertainment via Nokia's Ovi Life Tools, as well as the email account for the developing world, Ovi Mail.[111]
- **Nokia C2-01** gives the user the freedom to get the latest news,[112] check e-mails and keep up with social networks quickly and easily with a high-speed 3G connection.[88]
- **Nokia C2-02**[113] is a Touch and Type offering from Nokia, with a single SIM card slot and a sliding form factor. The phone includes a re-hashed Nokia Browser (formerly Ovi Browser) and an offline-available Series 40 version of Nokia Maps.[114]
- **Nokia C2-03**[115] is a dual-SIM variant of the Nokia C2-02, offering the same feature set with the addition of Easy Swap.[116]
- **Nokia C2-05** comes with the all-new Nokia Browser and Nokia Store, enabling seamless integration of online activities. The phone runs on Series 40, and comes with a VGA camera with digital zoom and full-screen viewfinder.[117]
- **Nokia C2-06** is a dual-SIM phone with attractive colours and high-quality materials that's easy to use – scroll through websites, photos and music on the 2.6" touch screen. Plus there's a slide-out keypad for fast, easy typing.[118] The phone also uses the swiping feature similar to that of Nokia N9
- **Nokia C3**[119] aims to get you online through a fast Wi-Fi connection, stay in touch with your friends, and share what's new through your social networks – all from the customizable home screen.
- **Nokia C3-01**[120] lets you flip through photos or browse Web pages on the touch screen and type using the keypad for easy messaging – all with one hand. Everything you need to stay in touch – right at your fingertips – with support for your favourite social networking, e-mail and chat services. The phone feels as good as it looks with its compact, sleek design and high-quality stainless steel casing.
- **Nokia C5**[121] boasts a cool feature in its phone book, where you can see your friends status updates directly from Facebook. You can also update your Facebook status easily and share your location with Ovi Maps 3.0.[122] The phone comes with a high-quality 3.2-megapixel camera.
- **Nokia C5 5MP**[123] is the latest release of the Nokia C5 smartphone. It offers the same feature set and appearance as its original version, but offers more Ram and a 5-megapixel camera, which allows for clearer pictures.[124]
- **Nokia C5-01** is a phone for TD-SCDMA. It runs S60 3rd Edition Feature Pack 2 on Symbian OS 9.3 and features a 5-megapixel camera and stainless steel materials.[125] It ships with support for a number of services including 139 Mailbox, Fetion IM, MM-Ovi Store and a range of pre-loaded content.[126]
- **Nokia C5-02** is China Telecom's offering with 2.6-inch QVGA resolution 260,000 color TFT screen.[127][128]
- **Nokia C5-03**[129] is one of the last few releases of Nokia under the Symbian^1 operating system as they move to Symbian^3, Symbian Anna and further releases.[130] It is also a budget resistive touchscreen smartphone with WLAN from the Cseries that was released in December 2010. The phone has Shazam music identification software and social networking services included or available for download.
- **Nokia C5-04**[131] is a WIND Mobile specific device featuring resistive touchscreen running on Symbian^1 operating system with HD voice. It also features a 2-megapixel camera.
- **Nokia C5-05**[132] is a downgraded variant of the Nokia C5-03, more for younger users. The features were retained, except for the camera which is downgraded to 2 megapixels, similar to the Canadian version.[133]
- **Nokia C5-06**[134] is a device similar to the Nokia C5-03, but lacks WLAN connectivity.
- **Nokia C5-08**[135] is another variant of Nokia C5-03, subsidized to European networks.[136]

- **Nokia C6**[1137] is a device with a similar form factor to Nokia N900 and Nokia N810. With a 3.2-inch HD touchscreen display and a slide out four-row QWERTY keyboard, the Nokia C6 brings the best of messaging together with instant access to your favorite contacts and communications. Also with 5-megapixel camera.
- **Nokia C6-01**[138] is a fully functional communicator with support for 3G and Wi-Fi onboard. It also has a 3.2-inch AMOLED capacative touch-screen ClearBlack display. The Nokia C6-01 is also one of Nokia's greenest devices to date, featuring the use of bioplastics and recycled metals for much of its construction.[139]
- **Nokia C7**[140] offers a full-touch glass display with social network integration, thousands of app offerings from the Ovi Store, customizable homescreens and 8-megapixel camera with HD video recording.[141]
- **Nokia C7 Astound**[142] is T-Mobile USA's version of Nokia C7 offering the same features, with the sudden addition of Wi-Fi calling.[143]
- **Nokia C7 Oro**[144][145] is a Nokia C7 made entirely of three luxurious and precious materials—sapphire crystal, leather and 18-carat gold. It also features an 8-megapixel high definition camera plus an award-winning headset, the Nokia J, which is a unique wireless headset with superior sound quality. It comes pre-loaded with a specially-customized Symbian Anna operating system, giving a theme that blends with the color of the handset. The phone will be made available only in select markets.[146]
- **Nokia C100** is a basic cellular phone featuring FM radio, flashlight, calendar, alarm clock and a grid-based menu system.[147]
- **Nokia C603** is a Symbian Belle device with a seamless, colorful and smooth design for the lower end of the market. The OS is offering up to six home screens and an extended range of widgets that can be placed in any combination the user chooses. The browser is four times faster than that found on previous Symbian versions, and offers enhanced compatibility with modern Web standards.[148]
- **Nokia C700**[149] is the smallest smartphone at the time of its launch.[150] It carries a 3.2" AMOLED display, NFC capabilities, and the first device to use the Symbian Belle operating system update. Home screens are designed to be more personalized,[151] often compared to Android's user interface.[149]
- **Nokia C701**[152] offers a high-resolution display using Nokia's proprietary ClearBlack technology[152] with IPS LCD.[153] It is one of the first three devices to carry the Symbian Belle operating system, and is armed with a 1 GHz processor. The phone supports embedded NFC, customizable widgets, informative lock screen and is similar to Nokia C7 in appearance.[154]
- **Nokia C1000** is a Chinese-specific mobile phone, for the lower part of the market. It provides different information about agriculture, lifestyle, education and weather through Ovi Life Tools. It runs on Series 30, and can play music through its dedicated MP3 player.
- **Nokia C1010** is a China-only mobile phone, offering dual-SIM ability and Easy Swap. It is made for emerging cities and provinces, where mobile phones are usually shared within a family or community. It offers MP3 player, Pinyin compatibility and Ovi Life Tools. It runs on the Series 30 operating system.

List of Zseries devices

Nokia Zseries

Manufacturer	Nokia
Series	Zseries
Availability by country	2010
Predecessor	Nokia Internet Tablet
Operating system	MeeGo
Ringtones & notifications	Nokia tune
Development status	Cancelled or discontinued

Nokia Zseries (stands for *Zuper*) is supposed to be Nokia's MeeGo tablet series. Originally intended to be released in 2010 and early 2011, the tablets were cancelled for unknown reasons. Only a product name has been revealed from the Ovi Store in November 2010.[155]

The list may also include upcoming devices or others that are previously intended to be a part of the series, but was scrapped or renamed due to Nokia's number-only naming change. These are the tablets included in the series:

- **Nokia Z500**[155] is the sole representative of the series, at least as leaked, and was shown on the Ovi Store in November 2010.[156] It was removed a few days later, leading speculations that the tablet will be postponed. The tablet runs on a native version of the MeeGo operating system, and is packed with an ARM chip.[157] On a Disney film called *Tron: Legacy*, the tablet was shown being used during a meeting in ENCOM Inc., by one of the movie's characters.[158] However, Eldar Murtazin tweeted that the tablet was to be cancelled due to the falling off of agreements between the carrier's pricing of the device, particularly AT&T.[159] Several demo videos appeared on video-sharing site YouTube, containing details on how the device works and how to use it.[160] Only one picture of the device has been known to exist as of the moment, and was captured inside Nokia's mobile development office by an unknown employee.[161] Patent applications for the said device have been gathered from the FCC files on March 2011, and since then, nothing was heard from Nokia Z500 anymore.[162]

List of Tseries devices

Nokia Tseries

Manufacturer	Nokia
Series	Tseries
Availability by country	2011—present
Operating system	Symbian
Ringtones & notifications	Nokia tune
Development status	Active

Nokia Tseries (stands for *Time*) is a China-specific line of phones, made to cater services offered in China where most technologies and Web features are disabled or replaced by the country's proprietary inventions. For instance, most of the phones offered in this series would include the TD-SCDMA and EV-DO networks which are dominant in that marketplace. The line was launched on 13 June 2011 in the Chinese version of the Nokia Conversations blog.[8]

The list may also include upcoming devices. These are the phones included in the series:

- **Nokia T7** is the first smartphone offered in the Tseries line, featuring a Nokia N8 shell with less specifications.[163] Instead of a 12-megapixel shooter, the camera was modified to only 8 megapixels, but still offers high quality picture-taking and HD video recording. It is notable that the HDMI is now gone, and the network that the phone supports are TD-SCDMA and GSM.[164]

- **Nokia T702** is a China-only handset, similar to the design of the Nokia E6 but with larger keys and apps suitable for the country. It also has deep integration of Chinese characters and special networking sites intended for that region. It is available to China Mobile only and will be offered in a modified Symbian Anna operating system.[165]

List of Aseries devices

Nokia Aseries

Manufacturer	Nokia
Series	Aseries
Availability by country	2011—present
Operating system	Series 40
Ringtones & notifications	Nokia tune
Development status	Active

The **Nokia Aseries** (stands for *Asha*, which means *hope*),[166] is a feature-phone line introduced by Nokia during Nokia World 2011.[167] The phones in the series are for developing markets, and more for younger users, for messaging and connectivity.

The list may also include upcoming devices. These are the phones included in the series:

- **Nokia A200** is a QWERTY featurephone, with dual-SIM capability and Easy Swap, enabling the user to change SIM cards without having to turn off the device. SIM manager has also been added in order to manage different profiles for different SIM cards inserted on the phone.[168]
- **Nokia A201** is a single SIM version of the aforementioned device. It is for music fans, with listening time of up to 7 hours and storage capacity of 32 GB.[167]
- **Nokia A300** is Nokia's Series 40 offering that combines a keypad and a touchscreen. The phone has 1GHz processor, 3G connectivity and preloaded with *Angry Birds*, the first time the game has been offered on non-smartphones.[167]
- **Nokia A303**[169] is a phone with QWERTY keypad input and a touchscreen. Armed with 1GHz processor, upgraded Nokia Browser, 2.6-inch touchscreen, and Nokia Store. Apps are readily available for download. The phone is said to break the line between featurephones and smartphones.[167]
- **Nokia A2000** is a China-specific featurephone with a QWERTY keypad. It supports storage of up to 32 GB through the use of microSD cards.
- **Nokia A2010** is a Chinese-only phone modeled after Nokia Asha 201 (or A201) featuring a QWERTY keypad and Asian language support.
- **Nokia A3000** is a touchscreen and keypad phone, for China only. It is modeled after the Nokia Asha 300 (or A300). The only difference found is the inclusion of Chinese language support.
- **Nokia A3030** is a touchscreen and QWERTY keypad phone, for Chinese countries. It is modeled after Nokia Asha 303 (or A303). Aside from the Chinese language support, it also has compatibility with *Angry Birds*, a 1 GHz processor and dedicated music buttons.

List of Lseries devices

Nokia Lseries

Manufacturer	Nokia
Series	Lseries
Availability by country	2011—present
Operating system	Windows Phone
Ringtones & notifications	Nokia tune
Development status	Active

The **Nokia Lseries** (stands for *Lumia*, which means *light*) is a series of Nokia's Windows Phone offerings, officially unveiled at the Nokia World 2011. This was created through an exclusive partnership in February of the same year, that will allow Nokia to use and modify Microsoft's mobile operating system.[170]

The phones included in the series are on this list:

- **Nokia L710**[170] is a Windows Phone-based device with a 3.7-inch capacitive touchscreen display and 1.4 GHz processor. The phone's back cover is interchangeable with a wide range of available colors. It also offers one-touch access to social networking sites, easy grouping of contacts, integrated communication threads and Internet Explorer 9.[171]
- **Nokia L800**[172] is a Windows Phone-based device with 1.4 GHz of processor with hardware acceleration and separate graphics processor, 3.7-inch AMOLED display with ClearBlack technology and 8-megapixel camera with Carl Zeiss optics.[10] It has 16 GB of internal memory, with 25 GB free from SkyDrive,[173] and uses Internet Explorer as its primary Web browser.[174] The phone is termed the world's "first real Windows Phone."[175] It comes with a host of Nokia-exclusive services like Nokia Drive, ESPN Sports Hub, Nokia Pulse[176] and Mix Radio from Nokia Music.[177] An accompanying pair of headphones were released with it too, named Nokia Purity HD Stereo Headset by Monster.[178]

References

[1] Rudy De Waele (29 April 2005). "Nokia Nseries planned 4th quarter 2005 | mTrends – mobile media lifestyle trends" (http://www.m-trends.org/2005/04/nokia-nseries-planned-4th-quarter-2005.html). m-trends.org. . Retrieved 12 November 2011.

[2] "Eseries Games" (http://web.archive.org/web/20101127143722/http://nokiamobileblog.com/category/nokia-eseries-mobile-devices/download-nokia-eseries-games-free/). Nokia Mobile Blog. 2010. Archived from the original (http://nokiamobileblog.com/category/nokia-eseries-mobile-devices/download-nokia-eseries-games-free/) on 2010-11-27. .

[3] "The Nokia E Series Range of Smartphones" (http://www.brighthub.com/mobile/symbian-platform/articles/88622.aspx). Brighthub.com. 27 September 2010. . Retrieved 12 November 2011.

[4] 6 Mar 2010 (6 March 2010). "Nokia's New Naming Convention: Does it make any C-E-N-X to you? | NokiaWorld" (http://symbianworld.org/3543-nokias-new-naming-convention-does-it-make-any-c-e-n-x-to-you/). Symbianworld.org. . Retrieved 12 November 2011.

[5] "at" (http://www.nampblog.com/2009/07/nokia-c-series-and-x-series-coming-at.html). Nampblog.com. . Retrieved 12 November 2011.

[6] "Nokia Xseries debuts with the X6 (Alvin) and X3 music phones" (http://www.unwiredview.com/2009/09/02/nokia-xseries-debuts-with-the-x6-alvin-and-x3-music-phones/). Unwired View. . Retrieved 12 November 2011.

[7] February, Shahbaaz (5 February 2010). "Nokia C Series Family : Nokia C3, C5 and C6 (Nokia Mystic) Heading For MWC Announcement?" (http://tnerd.com/2010/02/05/nokia-c-series-family-nokia-c3-c5-and-c6-nokia-mystic-heading-for-mwc-announcement/). Tnerd.com. . Retrieved 12 November 2011.

[8] "□ □ TD-SCDMA□ □ □ □ □ □ □ □ □ T7-00□ □ □ □ 702T□ □ □ □ | □ □ □ □ □ " (http://blog.nokia.com.cn/2011/06/13/td-scdma-t700-702t/) (in Chinese). *blog.nokia.com.cn*. Nokia. 13 June 2011. . Retrieved 12 November 2011.

[9] Bartlett, Karen (26 October 2011). "Nokia's new Series 40 Asha range" (http://conversations.nokia.com/2011/10/26/nokiaâs-new-series-40-asha-range/). Nokia. . Retrieved 12 November 2011.

[10] Delaney, Ian (26 October 2011). "Lumia eclipse" (http://conversations.nokia.com/2011/10/26/lumia-eclipse/). Nokia. . Retrieved 12 November 2011.

[11] http://www.n-gage.com/

[12] http://xpressmusic.wordpress.com/about/
[13] Sorrel, Charlie (2007-08-07). "Nokia Prism Series: Diamonds are Forever" (http://www.wired.com/gadgetlab/2007/08/nokia-prism-ser/). *Gadget Lab* (Wired). . Retrieved 2012-01-23.
[14] Ziegler, Chris (2008-06-27). "Nokia unleashes Supernova series: meet the 7210, 7310, 7510, and 7610" (http://www.engadget.com/2008/06/27/nokia-unleashes-supernova-series-meet-the-7210-7310-7510-and/). Engadget. . Retrieved 2012-01-23.
[15] Bradshaw, Bryan (2010-08-11). "Nokia Fashion Collections, the Distinctly Bold Collection" (http://www.cell-phone-plans.net/blog/cell-phones/nokia-fashion-collections-the-distinctly-bold-collection/). *cell-phone-plans.net*. BHB Web Tech Ent.. . Retrieved 2012-01-23.
[16] "Nokia's Latest Fashion Collection: Oh L'Amour" (http://www.nokia.com/NOKIA_COM_1/About_Nokia/Press/Press_Events/zz_lamour/lamour.html) (Press release). Nokia. 2005-10-19. . Retrieved 2012-01-23.
[17] http://europe.nokia.com/find-products/nseries
[18] "Nokia N70 product page" (http://www.nseries.com/index.html#l=products,n70). Nseries.com. . Retrieved 12 November 2011.
[19] "Slide and Shoot with the Nokia N70: the smallest 3G Series 60 smartphone with 2 megapixel camera" (http://web.archive.org/web/20090210022714/http://press.nokia.com/PR/200504/991482_5.html) (Press release). Nokia. 2005-04-27. Archived from the original (http://press.nokia.com/PR/200504/991482_5.html) on 2009-02-10. .
[20] German, Kent (2006-05-05). "Nokia N-series line show" (http://reviews.cnet.com/4520-3504_7-6217771-1.html?tag=txt). *CNET Reviews*. clnet. . Retrieved 2012-01-23.
[21] "Nokia N71 product page" (http://www.nseries.com/index.html#l=products,n71). Nseries.com. . Retrieved 12 November 2011.
[22] "Nokia N71 – NokiaN71" (http://feetup.org/wiki/NokiaN71). Feetup.org. . Retrieved 12 November 2011.
[23] "The Nokia Experience – With Nokia N71, communication is clear" (http://thenokiaexperience.tumblr.com/post/4684079082/with-nokia-n71-communication-is-clear). Thenokiaexperience.tumblr.com. . Retrieved 12 November 2011.
[24] "Nokia N71 – Full phone specifications" (http://www.gsmarena.com/nokia_n71-1345.php). Gsmarena.com. . Retrieved 12 November 2011.
[25] "Nokia N72 product page" (http://www.nseries.com/index.html#l=products,n72). Nseries.com. . Retrieved 12 November 2011.
[26] "Nokia N72 Gloss Black (unlocked) Accessories" (http://www.mobilecityonline.com/wireless/store/productaccessory.asp?productid=21301). mobilecityONLINE.com. 2011. . Retrieved 2012-01-23.
[27] "Nokia N72 – Full phone specifications" (http://www.gsmarena.com/nokia_n72-1552.php). Gsmarena.com. . Retrieved 12 November 2011.
[28] "The Nokia Experience – Stylish covers and awesome hardware: What more can you ask for? The Nokia N72 shouts" (http://thenokiaexperience.tumblr.com/post/5827828523/stylish-covers-and-awesome-hardware-what-more-can-you-as). Thenokiaexperience.tumblr.com. . Retrieved 12 November 2011.
[29] "Nokia N73 product page" (http://www.nseries.com/index.html#l=products,n73). Nseries.com. . Retrieved 12 November 2011.
[30] http://ibnlive.in.com/photogallery/5200-6.html
[31] "Nokia N73 – Full phone specifications" (http://www.gsmarena.com/nokia_n73-1550.php). Gsmarena.com. 7 November 2011. . Retrieved 12 November 2011.
[32] "The Nokia Experience – The first phone to get the "multimedia computer" title: Meet the Nokia N73" (http://thenokiaexperience.tumblr.com/post/5857583228/the-first-phone-to-get-the-multimedia-computer-title-mee). thenokiaexperience.tumblr.com. . Retrieved 12 November 2011.
[33] Cha, Bonnie; German, Kent (2006-08-28). "Nokia N73 Review" (http://reviews.cnet.com/smartphones/nokia-n73-unlocked/4505-6452_7-31861312.html). *CNET Editors' Review*. clnet. . Retrieved 2012-01-23.
[34] "Nokia E50 product page" (http://www.europe.nokia.com/nokia/0,,91889,00.html). Nokia. . Retrieved 12 November 2011.
[35] (http://press.nokia.com/PR/200605/1051798_5.html)
[36] "Nokia Europe – Nokia E50 – Support" (http://europe.nokia.com/A4158064). Nokia. . Retrieved 12 November 2011.
[37] Nokia E51 product page (http://europe.nokia.com/A4546299)
[38] "Nokia E51 Review – Phones – CNET Asia" (http://asia.cnet.com/product/nokia-e51-41199918.htm). Asia.cnet.com. . Retrieved 12 November 2011.
[39] "Nokia E55 product page" (http://europe.nokia.com/find-products/devices/nokia-e55). Nokia. . Retrieved 12 November 2011.
[40] "Nokia E60 product page" (http://www.europe.nokia.com/nokia/0,,81338,00.html). Nokia. . Retrieved 12 November 2011.
[41] "Nokia E61product page" (http://europe.nokia.com/A4142101). Nokia. . Retrieved 12 November 2011.
[42] "Nokia E61i product page" (http://europe.nokia.com/A4344018). Nokia. . Retrieved 12 November 2011.
[43] "Nokia E61i – Full phone specifications" (http://www.gsmarena.com/nokia_e61i-1858.php). Gsmarena.com. . Retrieved 12 November 2011.
[44] "Nokia E62 product page" (http://usa.nokia.com/A4197015). Nokia. . Retrieved 12 November 2011.
[45] "Nokia E62 (finally) hits Cingular" (http://www.engadget.com/2006/09/12/nokia-e62-finally-hits-cingular/). Engadget. . Retrieved 12 November 2011.
[46] eegb81. "Nokia E62 Review – Watch CNET's Video Review" (http://reviews.cnet.com/smartphones/nokia-e62-at-t/4505-6452_7-32028401.html). Reviews.cnet.com. . Retrieved 12 November 2011.
[47] "Nokia E63 product page" (http://europe.nokia.com/find-products/devices/nokia-e63). Nokia. . Retrieved 12 November 2011.
[48] stevenj2. "Nokia E63 Review – Watch CNET's Video Review" (http://reviews.cnet.com/smartphones/nokia-e63-red-unlocked/4505-6452_7-33490444.html#ixzz1PMkQnKUq). Reviews.cnet.com. . Retrieved 12 November 2011.

[49] stevenj2. "Nokia E63 Review – Watch CNET's Video Review" (http://reviews.cnet.com/smartphones/nokia-e63-red-unlocked/4505-6452_7-33490444.html#ixzz1PMkIljgm). Reviews.cnet.com. . Retrieved 12 November 2011.

[50] eegb81. "Nokia E62 Review – Watch CNET's Video Review" (http://reviews.cnet.com/smartphones/nokia-e62-at-t/4505-6452_7-32028401.html#ixzz1PMjkbRRc). Reviews.cnet.com. . Retrieved 12 November 2011.

[51] stevenj2. "Nokia E63 Review – Watch CNET's Video Review" (http://reviews.cnet.com/smartphones/nokia-e63-red-unlocked/4505-6452_7-33490444.html?tag=mncol;lst;1). Reviews.cnet.com. . Retrieved 12 November 2011.

[52] "Nokia E65 product page" (http://europe.nokia.com/A4344227). Nokia. . Retrieved 12 November 2011.

[53] andy_taipei. "Nokia E65 Review – Watch CNET's Video Review" (http://reviews.cnet.com/smartphones/nokia-e65-unlocked/4505-6452_7-32329109.html?tag=mncol;lst;1). Reviews.cnet.com. . Retrieved 12 November 2011.

[54] E66 product page (http://europe.nokia.com/A41148254Nokia)

[55] "Nokia E70 product page" (http://www.europe.nokia.com/nokia/0,,81003,00.html). Nokia. . Retrieved 12 November 2011.

[56] Nokia E70-2 product page (http://www.nokiausa.com/A4409015)

[57] "Nokia E71 product page" (http://europe.nokia.com/A41146122). Nokia. . Retrieved 12 November 2011.

[58] Reviewed by Andrew Lim on 24 June 2008 (24 June 2008). "Nokia E71 Review | Mobile Phones | CNET UK" (http://reviews.cnet.co.uk/mobiles/0,39030107,49297522,00.htm). Reviews.cnet.co.uk. . Retrieved 12 November 2011.

[59] "Mobile Choice's Best Smartphone of the year – News – Mobile Choice" (http://www.mobilechoiceuk.com/News/Mobile+Choice's+Best+Smartphone+of+the+year/2713). Mobilechoiceuk.com. . Retrieved 12 November 2011.

[60] 5310 XpressMusic (25 July 2008). "This Smartphone Means Business, Lets You Toggle to Personal Life | Product Reviews" (http://www.wired.com/reviews/product/nokia_e71). *Wired*. . Retrieved 12 November 2011.

[61] "Global Mobile Awards 2012" (http://www.globalmobileawards.com/awards/winners_archive.shtml). Globalmobileawards.com. . Retrieved 12 November 2011.

[62] "Nokia Cell Phones, Touch Screen Phones, Business Phones – Nokia – USA" (http://www.nokiausa.com/find-products/phones/nokiae71x). Nokiausa.com. . Retrieved 12 November 2011.

[63] Kharif, Olga (30 March 2009). "Nokia Targets the U.S. Market" (http://www.businessweek.com/technology/content/mar2009/tc20090329_098616.htm). *BusinessWeek*. . Retrieved 12 November 2011.

[64] brunos82. "Nokia E71x Review – Watch CNET's Video Review" (http://reviews.cnet.com/smartphones/nokia-e71x-black-at/4505-6452_7-33573216.html?tag=contentMain;contentBody;1r). Reviews.cnet.com. . Retrieved 12 November 2011.

[65] "Nokia E72 product page" (http://europe.nokia.com/find-products/devices/nokia-e72). Nokia. . Retrieved 12 November 2011.

[66] "Nokia Europe – Nokia E72 – Mobile phone for business and personal communication with free navigation" (http://webcache.googleusercontent.com/search?q=cache:WHeeyHtr9CcJ:europe.nokia.com/find-products/devices/nokia-e72+Nokia+E72&cd=2&hl=tl&ct=clnk&gl=ph&source=www.google.com.ph). Webcache.googleusercontent.com. . Retrieved 12 November 2011.

[67] "Nokia Europe – Nokia E72 – White Edition business phone" (http://europe.nokia.com/find-products/devices/nokia-e72/white-edition). Nokia. . Retrieved 12 November 2011.

[68] "Nokia E72i – an E72 with WAPI for China?" (http://www.unwiredview.com/2010/03/09/nokia-e72i-an-e72-with-wapi-for-china/). Unwired View. . Retrieved 12 November 2011.

[69] "Nokia E73 Mode product page" (http://www.nokiausa.com/find-products/phones/nokia-e73). Nokiausa.com. . Retrieved 12 November 2011.

[70] "Nokia E73 Mode Review – Watch CNET's Video Review" (http://reviews.cnet.com/smartphones/nokia-e73-mode-t/4505-6452_7-34117965.html?tag=mncol;lst;1). Reviews.cnet.com. . Retrieved 12 November 2011.

[71] Cha, Bonnie (3 June 2010). "Affordable Nokia E73 Mode headed to T-Mobile | Dialed In – CNET Blogs" (http://www.cnet.com/8301-17918_1-20006691-85.html). Cnet.com. . Retrieved 12 November 2011.

[72] "Nokia E75 product page" (http://europe.nokia.com/e75#/main/landing). Nokia. . Retrieved 12 November 2011.

[73] "Nokia E75 Review – Watch CNET's Video Review" (http://reviews.cnet.com/smartphones/nokia-e75/4505-6452_7-33558906.html?tag=mncol;lst;1#reviewPage1). Reviews.cnet.com. . Retrieved 12 November 2011.

[74] "Nokia E90 product page" (http://europe.nokia.com/A4346040). Nokia. . Retrieved 12 November 2011.

[75] abdulla77. "Nokia E90 Communicator Review – Watch CNET's Video Review" (http://reviews.cnet.com/smartphones/nokia-e90-communicator/4505-6452_7-32329111.html#ixzz1PMukYbET). Reviews.cnet.com. . Retrieved 12 November 2011.

[76] abdulla77. "Nokia E90 Communicator Review – Watch CNET's Video Review" (http://reviews.cnet.com/smartphones/nokia-e90-communicator/4505-6452_7-32329111.html?tag=mncol;lst;1). Reviews.cnet.com. . Retrieved 12 November 2011.

[77] "Nokia E5 – Full phone specifications" (http://www.gsmarena.com/nokia_e5-3198.php). Gsmarena.com. . Retrieved 12 November 2011.

[78] mudphuck (18 October 2011). "Nokia E5-00 Review – Smartphones – CNET Reviews" (http://reviews.cnet.com/smartphones/nokia-e5-00-smartphone/4505-6452_7-34291264.html?tag=mncol;lst;1). Reviews.cnet.com. . Retrieved 12 November 2011.

[79] . "Nokia Europe – Nokia E6 smartphone with touch screen and QWERTY – Specifications" (http://europe.nokia.com/find-products/devices/nokia-e6-00/specifications). Nokia. . Retrieved 12 November 2011.

[80] Delaney, Ian (12 April 2011). "Launch: Nokia E6 – business is in hand" (http://conversations.nokia.com/2011/04/12/launch-nokia-e6/). Nokia. . Retrieved 12 November 2011.

[81] Whatley, James (14 September 2010). "Nokia E7 photo gallery" (http://conversations.nokia.com/2010/09/14/nokia-e7-photo-gallery/). Nokia. . Retrieved 12 November 2011.

[82] Whatley, James (14 September 2010). "Nokia expands Symbian^3 family with Nokia E7, Nokia C6-01 and Nokia C7" (http://conversations.nokia.com/2010/09/14/nokia-expands-symbian3-family-with-nokia-e7-nokia-c6-01-and-nokia-c7/). Nokia. . Retrieved 12 November 2011.

[83] Delaney, Ian (25 August 2011). "Launch: Nokia 100/101 – single and dual-SIM phones" (http://conversations.nokia.com/2011/08/25/launch-nokia-101100-single-and-dual-sim-phones/). Nokia. . Retrieved 12 November 2011.

[84] http://nokiaxseries.net/

[85] Delaney, Ian (8 March 2011). "Listen up! The Nokia X1-00 is here" (http://conversations.nokia.com/2011/03/08/listen-up-the-nokia-x1-00-is-here/). Nokia. . Retrieved 12 November 2011.

[86] Delaney, Ian (24 May 2011). "Twice as nice: Nokia X1-01 and Nokia C2-00 with Dual SIM" (http://conversations.nokia.com/2011/05/24/twice-as-nice-nokia-x1-01-and-nokia-c2-00-with-dual-sim-2/). Nokia. . Retrieved 12 November 2011.

[87] "Nokia X2 – Price, Specs, Release Date | TechPinas : Philippines' Technology News, Tips and Reviews Blog" (http://www.techpinas.com/2010/05/nokia-x2-price-specs-release-date.html). TechPinas. 2 May 2010. . Retrieved 12 November 2011.

[88] Delaney, Ian (22 November 2010). "Launch: Nokia C2-01 and Nokia X2-01" (http://conversations.nokia.com/2010/11/22/launch-nokia-c2-01-and-nokia-x2-01/). Nokia. . Retrieved 12 November 2011.

[89] http://dgui.files.wordpress.com/2011/07/nokia_x2-01x2-03_-_rm-709_doc_v3.pdf

[90] Whatley, James (2 September 2009). "Nokia X3 revealed" (http://conversations.nokia.com/2009/09/02/nokia-x3-revealed/). Nokia. . Retrieved 12 November 2011.

[91] Author nokiamania, 18.05.2010, 10:01 1 comment (18 May 2010). "Nokia X3 CDMA / EV-DO Version Released" (http://www.nokiaphones.net/nokia-x3-cdma-ev-do-version-released/). Nokiaphones.net. . Retrieved 12 November 2011.

[92] Delaney, Ian (17 August 2010). "Launch: Nokia X3 Touch and Type" (http://conversations.nokia.com/2010/08/17/launch-nokia-x3-touch-and-type/). Nokia. . Retrieved 12 November 2011.

[93] "Nokia X5 TD-SCDMA – Full phone specifications" (http://www.gsmarena.com/nokia_x5_td_scdma-3286.php). Gsmarena.com. . Retrieved 12 November 2011.

[94] "Nokia X5-00 and Nokia X5-01 Due To Launch?" (http://noknok.tv/2010/08/13/nokia-x5-00-and-nokia-x5-01-due-to-launch/). NokNok.tv. . Retrieved 12 November 2011.

[95] "Nokia X5-01" (http://www.youtube.com/watch?v=178itMFCE24). YouTube. . Retrieved 12 November 2011.

[96] Delaney, Ian (14 June 2010). "Hip to be Square: Two Xseries launches at Nokia Connection (photo gallery)" (http://conversations.nokia.com/2010/06/14/hip-to-be-square-two-xseries-launches-at-nokia-connection-photo-gallery/). Nokia. . Retrieved 12 November 2011.

[97] "Nokia Europe – Nokia X6" (http://europe.nokia.com/find-products/devices/nokia-x6). Nokia. . Retrieved 12 November 2011.

[98] Whatley, James (2 September 2009). "Nokia X6 launched" (http://conversations.nokia.com/2009/09/02/nokia-x6-launched/). Nokia. . Retrieved 12 November 2011.

[99] "Nokia Europe – Nokia X7 smartphone with a 4" touch screen – Overview" (http://europe.nokia.com/find-products/devices/nokia-x7-00). Nokia. . Retrieved 12 November 2011.

[100] Delaney, Ian (12 April 2011). "Launch: Nokia X7 – extreme entertainment" (http://conversations.nokia.com/2011/04/12/launch-nokia-x7/). Nokia. . Retrieved 12 November 2011.

[101] Segan, Sascha (1 August 2011). "Nokia Debuts 500 Phone, Changes Naming System | News & Opinion" (http://www.pcmag.com/article2/0,2817,2389499,00.asp). PCMag.com. . Retrieved 12 November 2011.

[102] Delaney, Ian (1 August 2011). "Launch: the Nokia 500 – fast, light and multicoloured" (http://conversations.nokia.com/2011/08/01/launch-the-nokia-500-fast-light-and-multicoloured/). Nokia. . Retrieved 12 November 2011.

[103] Delaney, Ian (24 August 2011). "Launch: Nokia 600 – loud and proud" (http://conversations.nokia.com/2011/08/24/launch-nokia-600-loud-and-proud/). Nokia. . Retrieved 12 November 2011.

[104] "Nokia 600 cancelled – will not be 'shipped to markets'" (http://www.allaboutsymbian.com/news/item/13535_Nokia_600_cancelled-will_not_b.php). All About Symbian. 2 November 2011. . Retrieved 12 November 2011.

[105] Whatley, James (2 March 2010). "Nokia naming conventions" (http://conversations.nokia.com/2010/03/02/nokia-naming-conventions/). Nokia. . Retrieved 12 November 2011.

[106] "Nokia Philippines – Nokia C1-00" (http://www.nokia.com.ph/find-products/products/nokia-c1-00). Nokia.com.ph. . Retrieved 12 November 2011.

[107] "Nokia Philippines – Nokia C1-01 – Products" (http://www.nokia.com.ph/find-products/products/nokia-c1-01). Nokia.com.ph. . Retrieved 12 November 2011.

[108] "Nokia Philippines – Nokia C1-02 – Products" (http://www.nokia.com.ph/find-products/products/nokia-c1-02). Nokia.com.ph. . Retrieved 12 November 2011.

[109] "Google Translate" (http://translate.google.com/translate?js=n&prev=_t&hl=en&ie=UTF-8&layout=2&eotf=1&sl=auto&tl=en&u=http://www.nokia.com.cn/find-products/products/c1-03). Google. . Retrieved 12 November 2011.

[110] "Nokia Philippines – Nokia C2-00 – Products" (http://www.nokia.com.ph/find-products/products/nokia-c2-00). Nokia.com.ph. 24 May 2011. . Retrieved 12 November 2011.

[111] Morgan, Rhiain (3 June 2010). "Dual SIM Nokia C2 revealed (photo gallery)" (http://conversations.nokia.com/2010/06/03/dual-sim-nokia-c2-revealed-photo-gallery/). Nokia. . Retrieved 12 November 2011.

[112] "Nokia Philippines – Nokia C2-01 3G mobile phone – Products" (http://www.nokia.com.ph/find-products/products/nokia-c2-01). Nokia.com.ph. . Retrieved 12 November 2011.

[113] "Nokia Europe – Nokia C2-02 Touch and Type mobile phone – Overview" (http://europe.nokia.com/find-products/devices/nokia-c2-02). Nokia. . Retrieved 12 November 2011.

[114] "Nokia C2-02, Nokia C2-03, Nokia C2-06 Touch and Type launched at Nokia Connection 2011" (http://noknok.tv/2011/06/21/nokia-c2-02-nokia-c2-03-nokia-c2-06-touch-and-type-launched-at-nokia-connection-2011/). NokNok.tv. . Retrieved 12 November 2011.

[115] "Nokia Europe – Nokia C2 Dual SIM Touch and Type – Overview" (http://europe.nokia.com/find-products/devices/nokia-c2-03). Nokia. . Retrieved 12 November 2011.

[116] "Nokia announces C2-02, C2-03 and C2-06 trio of feature phones – GSMArena.com news" (http://www.gsmarena.com/nokia_c203_dual_sim_feature_phone_is_announced-news-2804.php). Gsmarena.com. . Retrieved 12 November 2011.

[117] Delaney, Ian (11 October 2011). "Launch: Nokia C2-05 and Nokia X2-05 – looks and power, for less" (http://conversations.nokia.com/2011/10/11/launch-nokia-c2-05-and-nokia-x2-05-â-looks-and-power-for-less/). Nokia. . Retrieved 12 November 2011.

[118] "Phone Finder and Compare" (http://www.phonearena.com/phones/Nokia-C2-06_id5427http://www.phonearena.com/phones/Nokia-C2-06_id5427). Phonearena.com. . Retrieved 12 November 2011.

[119] "Nokia Philippines – Nokia C3-00 – Products" (http://www.nokia.com.ph/find-products/products/nokia-c3-00). Nokia.com.ph. . Retrieved 12 November 2011.

[120] "Nokia Philippines – Nokia C3 Touch and Type mobile phone – Products" (http://www.nokia.com.ph/find-products/products/nokia-c3-touch-and-type). Nokia.com.ph. . Retrieved 12 November 2011.

[121] "Nokia Philippines – Nokia C5-00 5MP camera phone – Products" (http://www.nokia.com.ph/find-products/products/nokia-c5-00). Nokia.com.ph. . Retrieved 12 November 2011.

[122] Whatley, James (2 March 2010). "Nokia C5 unveiled" (http://conversations.nokia.com/2010/03/02/nokia-c5-unveiled/). Nokia. . Retrieved 12 November 2011.

[123] "Nokia Europe – Nokia C5 00 5MP camera phone – Products" (http://europe.nokia.com/find-products/devices/nokia-c5-00). Nokia. . Retrieved 12 November 2011.

[124] "Nokia C5 5MP – Full phone specifications" (http://www.gsmarena.com/nokia_c5_5mp-4019.php). Gsmarena.com. . Retrieved 12 November 2011.

[125] "Nokia C5-01 – classically styled and affordable TD-SCDMA handset" (http://www.allaboutsymbian.com/news/item/11438_Nokia_C5-01_classically_styled.php). All About Symbian. 23 April 2010. . Retrieved 12 November 2011.

[126] "Mobile Themes for Nokia C5-01" (http://uithemes.com/en/For-Nokia-C5-01/). UI Themes. . Retrieved 12 November 2011.

[127] "C5-02 Nokia special custom first exposure _NOKIA Nokia" (http://www.waybeta.com/news/32708/c5-02-nokia-special-custom-first-exposure-_nokia-nokia/). waybeta. 9 August 2010. . Retrieved 12 November 2011.

[128] "Nokia C5-02 Specifications" (http://www.mobilerated.com/nokia-c5-02-specifications.html). Mobilerated.com. . Retrieved 12 November 2011.

[129] "Nokia Philippines – Nokia C5-03 touch screen phone – Products" (http://www.nokia.com.ph/find-products/products/nokia-c5-03). Nokia.com.ph. . Retrieved 12 November 2011.

[130] "Search Result" (http://conversations.nokia.com/search-result/?q=Nokia+C5-03&sa=Â &cx=016381219545844074384:4unqkykcbqu&cof=FORID:9;NB:1&ie=UTF-8#938). Nokia. . Retrieved 12 November 2011.

[131] "WIND Mobile | Shop | Phones | Nokia C5-04" (http://shop.windmobile.ca/ProductCatalog/Handsets/HandsetDetails.aspx?id=Nokia+C5(WINDCA)&color=orange). Shop.windmobile.ca. . Retrieved 12 November 2011.

[132] "Nokia Europe – Nokia C5-05 touch screen phone" (http://europe.nokia.com/find-products/devices/nokia-c5-05). Nokia. . Retrieved 12 November 2011.

[133] "Nokia C5-06 and C5-05 pop up, are cheaper versions of C5-03 – GSMArena.com news" (http://www.gsmarena.com/nokia_c506_and_c505_pop_up_are_cheaper_versions_of_c503-news-3246.php). Gsmarena.com. . Retrieved 12 November 2011.

[134] "Nokia Europe – Nokia C5-06 touchscreen phone with free navigation" (http://europe.nokia.com/find-products/devices/nokia-c5-06). Nokia. . Retrieved 12 November 2011.

[135] http://europe.nokia.com/PRODUCT_METADATA_0/Products/Phones/C-Series/C5-06/Images/c5-06_black_orange_front_604x604.png

[136] http://nds1.nokia.com/declaration_of_confirmity/files/declaration_of_conformity/Nokia_C5-08_-__RM-814_R&TTE_DoC.pdf

[137] Morgan, Rhiain (13 April 2010). "Nokia C6 unveiled (photo gallery)" (http://conversations.nokia.com/2010/04/13/nokia-c6-unveiled/). Nokia. . Retrieved 12 November 2011.

[138] "Nokia Philippines – Nokia C6 smartphone with a ClearBlack AMOLED touch screen – Overview" (http://www.nokia.com.ph/find-products/products/nokia-c6-01). Nokia.com.ph. . Retrieved 12 November 2011.

[139] Delaney, Ian (4 November 2010). "Nokia C6-01 starts shipping today" (http://conversations.nokia.com/2010/11/04/nokia-c6-01-starts-shipping-today/). Nokia. . Retrieved 12 November 2011.

[140] "Nokia Philippines – Nokia C7-00" (http://www.nokia.com.ph/find-products/products/nokia-c7-00). Nokia.com.ph. . Retrieved 12 November 2011.

[141] Schwarzmann, Phil (11 October 2010). "Nokia C7 is shipping!" (http://conversations.nokia.com/2010/10/11/nokia-c7-is-shipping/). Nokia. . Retrieved 12 November 2011.

[142] "Nokia Astound – Nokia – USA" (http://www.nokiausa.com/find-products/phones/nokia-c7-00?intc=ncomprod-fw-ilc-bdy-acq-na-nokiacom-us-20-c7ndphn_0x0). Nokiausa.com. . Retrieved 12 November 2011.

[143] "Nokia Cell Phones & Smartphones| Wirefly | Mobile Phone" (http://phones.nokiausa.com/eCommerce/SpecialOffer.aspx?cid=36285_b4d3486b3def4275ac66f0e926ba6324). Phones.nokiausa.com. . Retrieved 12 November 2011.

[144] "Nokia Europe – Nokia Oro touch screen smartphone with 18-carat gold plating – Overview" (http://europe.nokia.com/find-products/devices/nokia-oro). Nokia. . Retrieved 12 November 2011.

[145] http://nds1.nokia.com/declaration_of_confirmity/files/declaration_of_conformity/Nokia_C7-00s_-__RM-749_R&TTE_DoC.pdf

[146] Delaney, Ian (25 May 2011). "Introducing Nokia Oro" (http://conversations.nokia.com/2011/05/25/gold-ringer-introducing-nokia-oro/). Nokia. . Retrieved 12 November 2011.

[147] "Nokia Europe – Nokia 100 mobile phone – Features" (http://europe.nokia.com/find-products/devices/nokia-100/features). Nokia. . Retrieved 12 November 2011.

[148] Delaney, Ian (13 October 2011). "Nokia 603 is Belle-issimo!" (http://conversations.nokia.com/2011/10/13/belle-issimo-nokia-603-launched-today/). Nokia. . Retrieved 12 November 2011.

[149] "Nokia Europe – Nokia 700 ultra-slim smartphone with bright ClearBlack display – Overview" (http://europe.nokia.com/find-products/devices/nokia-700). Nokia. . Retrieved 12 November 2011.

[150] "Nokia 700 specs" (http://www.phonearena.com/phones/Nokia-700_id5669). Phonearena.com. . Retrieved 12 November 2011.

[151] Delaney, Ian (24 August 2011). "Launch: Nokia 700 – the smallest smartphone in the world" (http://conversations.nokia.com/2011/08/24/launch-nokia-700-the-smallest-smartphone-in-the-world). Nokia. . Retrieved 12 November 2011.

[152] "Nokia Europe – Nokia 701 smartphone with a 3.5-inch ClearBlack display – Overview" (http://europe.nokia.com/find-products/devices/nokia-701). Nokia. . Retrieved 12 November 2011.

[153] Delaney, Ian (24 August 2011). "Launch: Nokia 701 – the brightest screen in the world" (http://conversations.nokia.com/2011/08/24/launch-nokia-701-the-brightest-screen-in-the-world/). Nokia. . Retrieved 12 November 2011.

[154] Delaney, Ian (24 August 2011). "Symbian Belle – the facts, the features and the pictures" (http://conversations.nokia.com/2011/08/24/symbian-belle-the-facts-the-features-and-the-pictures/). Nokia. . Retrieved 12 November 2011.

[155] "Nokia Z500 MeeGo tablet leaked on Ovi Store?" (http://www.engadget.com/2010/11/01/nokia-z500-meego-tablet-leaked-on-ovi-store/). Engadget. 1 November 2010. . Retrieved 12 November 2011.

[156] "Nokia Z500 Tablet leaked by Ovi Store | Mobile Geek Inc" (http://mobilegeekinc.com/2010/10/31/nokia-z500-tablet-leaked-by-ovi-store/). 31 October 2010. . Retrieved 12 November 2011.

[157] "@EldarMurtazin's June tweet right about the Arm based Nokia Z500? – Nokia MeeGo tablet makes appearance on Ovi Store" (http://mynokiablog.com/2010/11/01/eldarmurtazins-june-tweet-right-about-the-arm-based-nokia-z500-nokia-meego-tablet-makes-appearance-on-ovi-store/). My Nokia Blog. . Retrieved 12 November 2011.

[158] "Nokia's MeeGo Tablet makes cameo in TRON: Legacy? #Z500" (http://mynokiablog.com/2011/01/09/nokias-meego-tablet-makes-cameo-in-tron-legacy/). My Nokia Blog. 21 January 2011. . Retrieved 12 November 2011.

[159] Nguyen, Chuong (30 November 2010). "Nokia Z500 MeeGo Tablet Rumored to be Delayed Because of Price" (http://www.gottabemobile.com/2010/11/30/nokia-z500-meego-tablet-rumored-to-be-delayed-because-of-price/). Gottabemobile.com. . Retrieved 12 November 2011.

[160] "Nokia Z500 – Nokia Tablet – Ipad Killer" (http://www.youtube.com/watch?v=QTgxteroJc8). YouTube. 2 November 2010. . Retrieved 12 November 2011.

[161] "Is this the Nokia Tablet Z500? » Nokia, MeeGo, Intel, Instead, Twitter, MWC2011" (http://smartphonesmobile.net/nokia/nokia-tablet-z500/). Smartphones Mobile. 24 January 2011. . Retrieved 12 November 2011.

[162] Nguyen, Chuong (16 March 2011). "Nokia Tablet Patents: Will We See a MeeGo or Symbian Tablet Soon?" (http://www.gottabemobile.com/2011/03/16/nokia-tablet-patents-will-we-see-a-meego-or-symbian-tablet-soon/). Gottabemobile.com. . Retrieved 12 November 2011.

[163] "Nokia China announce Symbian^3 powered T7-00 and 702T" (http://www.allaboutsymbian.com/news/item/12999_Nokia_China_announce_Symbian3_.php). All About Symbian. 14 June 2011. . Retrieved 12 November 2011.

[164] "Photos of Symbian^3-running Nokia T7-00 leak, some specs too – GSMArena.com news" (http://www.gsmarena.com/photos_of_the_rumored_nokia_t700_leak-news-2478.php). Gsmarena.com. . Retrieved 12 November 2011.

[165] Gadget Guru (22 June 2011). "nokia 702 | nokia 702 features | nokia phone" (http://living.oneindia.in/gadgets/mobile/nokia-702-features-220611-aid0161.html). Living.oneindia.in. . Retrieved 12 November 2011.

[166] "5 things you didn't know about the Nokia Asha Mobile Phones" (http://nokiaconnects.com/2011/10/27/5-things-you-didnt-know-about-the-nokia-asha-mobile-phones/). Nokia Connects. 10 October 2011. . Retrieved 12 November 2011.

[167] Bartlett, Karen (26 October 2011). "Nokia's new Series 40 Asha range" (http://conversations.nokia.com/2011/10/26/nokiaâs-new-series-40-asha-range/). Nokia. . Retrieved 12 November 2011.

[168] 26 October 2011 by Stan Schroeder 9 (26 October 2011). "Nokia Unveils Four New Series 40 Phones" (http://mashable.com/2011/10/26/nokia-four-symbian-smartphones/). Mashable.com. . Retrieved 12 November 2011.

[169] "Nokia Europe – Nokia Asha 303 with QWERTY keyboard and a 2.6-inch touch screen – Overview" (http://europe.nokia.com/find-products/devices/nokia-asha-303). Nokia. . Retrieved 12 November 2011.

[170] "Lumia 710 – Nokia – UK" (http://www.nokia.co.uk/gb-en/products/phone/lumia710/). Nokia. . Retrieved 12 November 2011.

[171] "Nokia Lumia 710 hands-on (video)" (http://www.engadget.com/2011/10/26/nokia-lumia-710-hands-on-video/). Engadget. 26 October 2011. . Retrieved 12 November 2011.

[172] "Lumia 800 – Nokia – UK" (http://www.nokia.co.uk/gb-en/products/phone/lumia800/). Nokia. . Retrieved 12 November 2011.

[173] "Nokia Lumia 800 shipping in November for $585, available for pre-order now" (http://www.engadget.com/2011/10/26/nokia-lumia-800-shipping-in-november-for-585-available-for-pre/). Engadget. 26 October 2011. . Retrieved 12 November 2011.
[174] "Nokia announces its Drive navigation, Mix Radio, and ESPN Sports Hub cloud services for WP7" (http://www.engadget.com/2011/10/26/nokia-announces-drive-mix-radio-cloud-services/). Engadget. 26 October 2011. . Retrieved 12 November 2011.
[175] "Nokia Lumia 800 hands-on (video)" (http://www.engadget.com/2011/10/26/nokia-lumia-800-hands-on/). Engadget. 26 October 2011. . Retrieved 12 November 2011.
[176] "Nokia Pulse" (http://pulse.nokia.com/welcome). Nokia. . Retrieved 12 November 2011.
[177] 26 October 2011 by Stan Schroeder 22 (26 October 2011). "Lumia 800: Nokia's First Windows Phone" (http://mashable.com/2011/10/26/lumia-800/). Mashable.com. . Retrieved 12 November 2011.
[178] "Nokia unveils Purity HD Stereo Headset with a little help from Monster" (http://www.engadget.com/2011/10/26/nokia-unveils-purity-hd-stereo-headset-with-a-little-help-from-m/). Engadget. 26 October 2011. . Retrieved 12 November 2011.

Mobile_operating_system

A **mobile operating system** (mobile OS) is the operating system that controls a smartphone, tablet, PDA, or other mobile device. Modern mobile operating systems combine the features of a personal computer operating system with touchscreen, cellular, Bluetooth, WiFi, GPS mobile navigation, camera, video camera, speech recognition, voice recorder, music player, Near field communication, personal digital assistant (PDA), and other features.

History

Mobile operating system milestones mirror the development of mobile phones and smartphones:

- 1979–1992 Mobile phones have embedded systems to control operation.
- 1993 The first smartphone, the IBM Simon, had a touchscreen, email, and PDA features.
- 1996 Palm Pilot 1000 personal digital assistant is introduced with the Palm OS mobile operating system.
- 1996 First Windows CE Handheld PC devices are introduced.
- 1999 Nokia S40 OS was officially introduced with the launch of the Nokia 7110
- 2000 Symbian became the first modern mobile OS on a smartphone with the launch of the Ericsson R380.
- 2001 The Kyocera 6035 is the first smartphone with Palm OS.
- 2002 Microsoft's first Windows CE (Pocket PC) smartphones are introduced.
- 2002 BlackBerry releases its first smartphone.
- 2005 Nokia introduced Maemo OS on the first internet tablet N770.
- 2007 Apple iPhone with iOS introduced as an iPod, "mobile phone" and "internet communicator."[1]
- 2007 Open Handset Alliance (OHA) formed by Google, HTC, Sony, Dell, Intel, Motorola, Samsung, LG, etc.[2]
- 2008 OHA releases Android 1.0 with the HTC Dream (T-Mobile G1) as the first Android phone.
- 2009 Palm introduced webOS with the Palm Pre. By 2012 webOS devices were no longer sold.
- 2009 Samsung announces the Bada OS with the introduction of the Samsung S8500.
- 2010 Windows Phone OS phones are released but are not compatible with the previous Windows Mobile OS.
- 2011 The MeeGo the first mobile Linux, combined Maemo and Moblin, was introduced with Nokia N9 in effect of cooperation of Nokia, Intel and Linux Foundation
- In September 2011 Intel and the Linux Foundation announced that their efforts will shift from MeeGo to Tizen during 2011 and 2012.
- In October 2011 the Mer project was announced, centered around a ultra-portable Linux + HTML5/QML/JS Core for building products with, derived from the MeeGo codebase.
- 2012 The Lenovo K800 will be the first Intel powered smartphone (Android OS).[3]

Market share

In 2006, Android, iOS, Windows Phone, and Bada did not yet exist and just 64 million smartphones were sold.[3] Today, nearly 10 times as many smartphones are sold and the top mobile operating systems marketed as "smartphones" by market share are Android, Symbian, Apple iOS, RIM BlackBerry, MeeGo, Windows Phone, and Bada.[4] Note that these statistics only include operating systems marketed as "smartphones" (for example, they don't include Nokia's S40 operating system, which according to Nokia's announcement on 25 January 2012, has sold over 1.5 billion S40 devices. [5]).

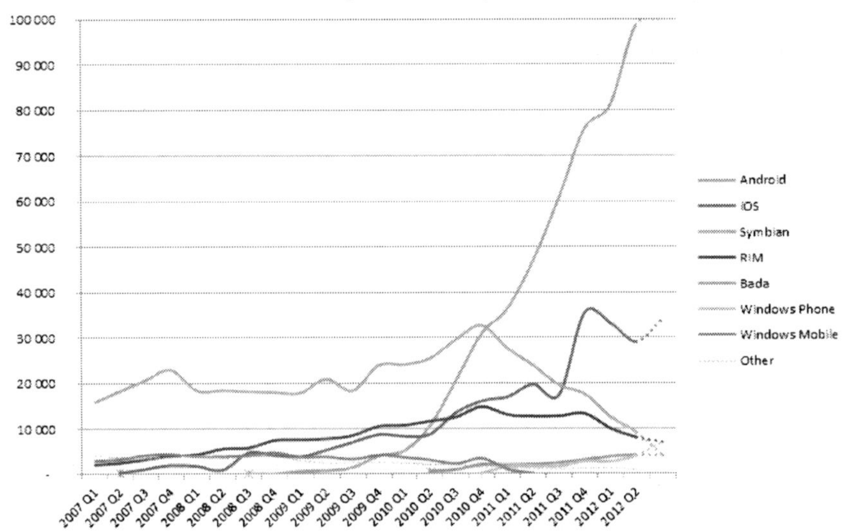

See table below for source data.

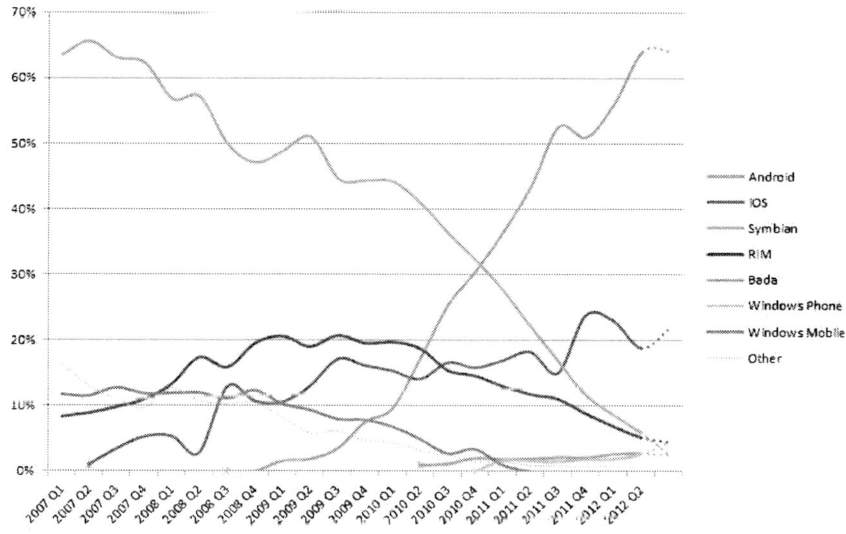

See table below for source data.

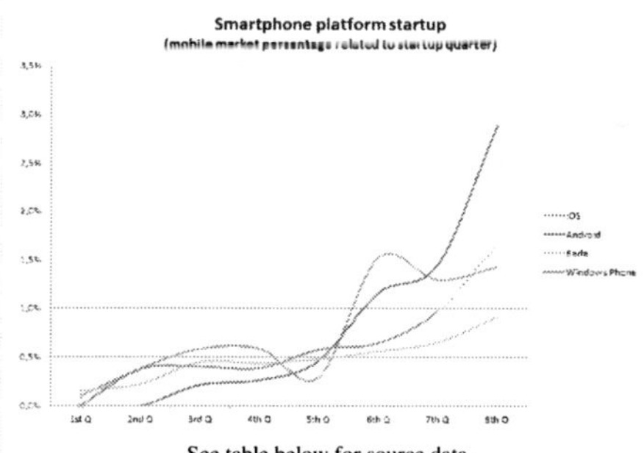

See table below for source data.

Gartner: World-Wide Smartphone Sales (Thousands of Units)

Quarter	Windows Mobile[6]	RIM	Symbian[7]	iOS	Android[8]	Bada	Windows Phone[9]	Other	Total Smartphones	Total
2012 Q2 [10]		7,991(5.2%/1.9%)	9,072(5.9%/2.2%)	28,935(18.8%/6.9%)	98,529(64.1%/23.5%)	4,209(2.7%/1.0%)	4,087(2.7%/1.0%)	863(0.6%/0.2%)	153,686(100%/36.7%)	419,008(1
2012 Q1 [11]		9,939(6.9%/2.4%)	12,467(8.6%/3.0%)	33,121(22.9%/7.9%)	81,067(56.1%/19.3%)	3,842(2.7%/0.9%)	2,713(1.9%/0.6%)	1 243(0.9%/0.3%)	144,392(100%/34.5%)	419,108(1
2011 Q4 [12]		13,185	17,458	35,456	75,906	3,111	2,759	1,167	149,042	476
2011 Q3 [4]		12,701	19,500	17,295	60,490	2,479	1,702	1,018	115,185	44
2011 Q2 [13]		12,652	23,853	19,629	46,776	2,056	1,724	1,051	107,740	428
2011 Q1 [11][14]	982	13,004	27,599	16,883	36,350	1,862	1,600	1,495	99,775	42
2010 Q4 [12]	3,419	14,762	32,642	16,011	30,801	2,027	0	1,488	101,150	45
2010 Q3 [4]	2,204	12,508	29,480	13,484	20,544	921		1,991	81,133	41
2010 Q2 [13]	3,059	11,629	25,387	8,743	10,653	577		2,011	62,058	36
2010 Q1 [14]	3,696	10,753	24,068	8,360	5,227			2,403	54,506	35
2009 Q4 [15]	4,203	10,508	23,857	8,676	4,043			2,517	53,804	34
2009 Q3 [16]	3,260	8,523	18,315	7,040	1,425			2,531	41,093	308
2009 Q2 [17]	3,830	7,782	20,881	5,325	756			2,398	40,972	28
2009 Q1 [18]	3,739	7,534	17,825	3,848	575			2,986	36,507	26
2008 Q4 [19]	4,714	7,443	17,949	4,079	0			3,958	38,143	31
2008 Q3 [20]	4,053	5,800	18,179	4,720	0			3,763	36,515	308
2008 Q2 [21]	3,874	5,594	18,405	893				3,456	32,221	30
2008 Q1 [19]	3,858	4,312	18,400	1,726				4,113	32,408	29
2007 Q4 [19]	4,374	4,025	22,903	1,928				3,536	36,766	33

4,180	3,192	20,664	1,104				3,612	32,752	291,142
3,212	2,471	18,273	270				3,628	27,855	272,604
2,931	2,080	15,844					4,087	24,943	259,039

Gartner: World-Wide Smartphone Sales (% of Smartphones / % of All phones)

Quarter	Windows Mobile	RIM	Symbian	iOS	Android	Bada	Windows Phone	Other Smartphones	Total Smartphones
2012 Q2		5.2%/1.9%	5.9%/2.2%	18.8%/6.9%	64.1%/23.5%	2.7%/1.0%	2.7%/1.0%	0.6%/0.2%	100.0%/36.7%
2012 Q1		6.9%/2.4%	8.6%/3.0%	22.9%/7.9%	56.1%/19.3%	2.7%/0.9%	1.9%/0.6%	0.9%/0.3%	100.0%/34.5%
2011	0.2%/0.1%	10.9%/2.9%	18.7%/5.0%	18.9%/5.0%	46.5%/12.4%	2.0%/0.5%	1.7%/0.4%	1.0%/0.3%	100.0%/26.6%
2010	4.1%/0.8%	16.6%/3.1%	37.3%/7.0%	15.6%/2.9%	22.5%/4.2%	1.2%/0.2%	0.0%/0.0%	2.6%/0.5%	100.0%/18.7%
2009	8.7%/1.2%	19.9%/2.8%	46.9%/6.7%	14.4%/2.1%	3.9%/0.6%			6.1%/0.9%	100.0%/14.2%
2008	11.8%/1.3%	16.6%/1.9%	52.4%/6.0%	8.2%/0.9%	0.0%/0.0%			11.0%/1.3%	100.0%/11.4%
2007	12.0%/1.3%	9.6%/1.0%	63.5%/6.7%	2.7%/0.3%				12.2%/1.3%	100.0%/10.6%

IDC: World-Wide Smartphone Shipments (Millions of Units)[22]

Quarter	Android[23]	iOS	Symbian[24]	BlackBerry OS	Linux[25]	Windows Phone	Other	Total
2012 Q2[26]	104.8	26.0	6.8	7.4	3.5	5.4	0.1	154.0
2012 Q1	89.9	35.1	10.4	9.7	3.5	3.3	0.4	152.3
2011 Q4	83.4	36.3	18.3	12.8	3.8	2.4	0.8	157.8
2011 Q3	67.7	16.3	17.3	11.3	3.9	1.4	0.1	118.1
2011 Q2[27]	50.8	20.4	18.3	12.5	3.3	2.5	0.6	108.4
2011 Q1	36.7	18.6	26.4	13.8	3.2	2.6	0.3	101.6

Current market share and outlook

IDC forecasts growth in market share of the Android from 59% in Q1 2012 to 61% in the entire year 2012 and to decline to 53% until 2016. The Windows Phone to grow from 2% in Q1 2012 to 5% in the entire year 2012 and then grow to 19% until 2016. The iPhone (iOS) to decline from 23% in Q1 2012 to 21% in the entire year 2012 and then decline to 19% until 2016. The BlackBerry market share is expected to keep current 6% until 2016.[28] Although, all forecasts are questionable, they do purportedly provide a good outlook to current and future trends.

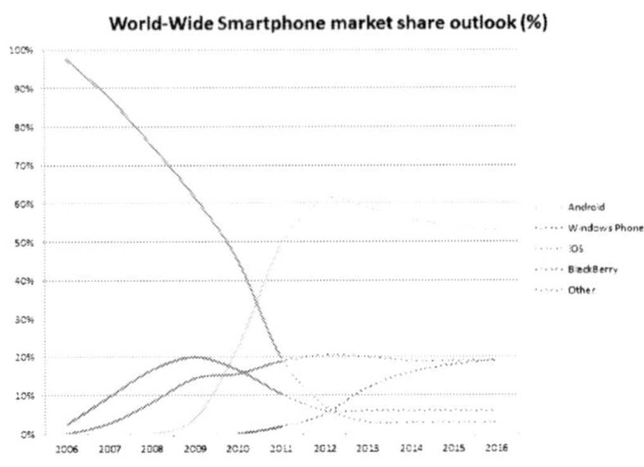

Based on IDC actuals and forecast

Mobile internet traffic share

As of June 2012, mobile data usage shows 65% of mobile data traffic to be from iOS, 20% from Android, 2% from BlackBerry, 1% from Symbian, and 1% from Windows Phone.[29]

Common software platforms

The most common mobile operating systems are:

Android from Google Inc.[30] (free and open source)[31]

Android was developed by a small startup company that was purchased by Google Inc. in 2005, and Google continues to update the software. Android is a Linux-derived OS backed by Google, along with major hardware and software developers (such as Intel, HTC, ARM, Samsung, Motorola and eBay, to name a few), that form the Open Handset Alliance.[32] Released on November 5th 2007, the OS received praise from a number of developers upon its introduction.[33] Android releases prior to 2.0 (1.0, 1.5, 1.6) were used exclusively on mobile phones. Most Android phones, and some Android tablets, now use a 2.x release. Android 3.0 was a tablet-oriented release and does not officially run on mobile phones. The current Android version is 4.0. Android releases are nicknamed after sweets or dessert items like Cupcake (1.5), Frozen Yogurt (2.2), Honeycomb (3.0), and Ice Cream Sandwich (4.0). Most major mobile service providers carry an Android device. Since the HTC Dream was introduced, there has been an explosion in the number of devices that carry Android OS. From Q2 of 2009 to the second quarter of 2010, Android's worldwide market share rose 850% from 1.8% to 17.2%. On 15 November 2011, Android reached 52.5% of the global smartphone market share.[34]

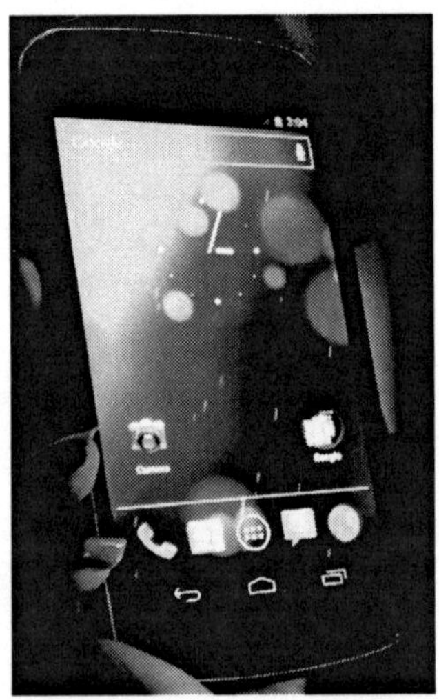

The Galaxy Nexus running Android 4.0.1

BlackBerry OS from RIM (closed source, proprietary)

This OS is focused on easy operation and was originally designed for business. Recently it has seen a surge in third-party applications and has been improved to offer full multimedia support. Currently Blackberry's App World has over 50,000 downloadable applications. RIM's future strategy will focus on the newly acquired QNX, having already launched the BlackBerry PlayBook tablet running a version of QNX and expecting the first QNX smartphones in early 2012.[35]

The Apple iPad tablet computer uses a version of iOS.

iOS from Apple Inc.[30] (closed source, proprietary, on top of open source Darwin core OS)

The Apple iPhone, iPod Touch, iPad, and second-generation Apple TV all use an operating system called iOS, which is derived from Mac OS X. Native third party applications were not officially supported until the release of iOS 2.0 on July 11th 2008. Before this, "jailbreaking" allowed third party applications to be installed, and this method is still available. Currently all iOS devices are developed by Apple and manufactured by Foxconn or another of Apple's partners.

S40 (Series40) from Nokia (closed source, proprietary)

Nokia uses S40 OS in their low end phones (aka feature phones). Over the years over 150 phone models have been developed running S40 OS [36]. Since the introduction of S40 OS it has evolved from monochrome low resolution UI to full touch 256k color UI.

Symbian OS from Nokia and Accenture[30] (open public license)

Symbian has the largest smartphone share in most markets worldwide, but lags behind other companies in the relatively small but highly visible North American market.[37] This matches the success of Nokia in all markets except Japan. In Japan Symbian is strong due to a relationship with NTT DoCoMo, with only one of the 44 Symbian handsets released in Japan coming from Nokia.[38] It has been used by many major handset manufacturers, including BenQ, Fujitsu, LG, Mitsubishi, Motorola, Nokia, Samsung, Sharp, and Sony Ericsson. Current Symbian-based devices are being made by Fujitsu, Nokia, Samsung, Sharp, and Sony Ericsson. Prior to 2009 Symbian supported multiple user interfaces, i.e. UIQ from UIQ Technologies, S60 from Nokia, and MOAP from NTT DOCOMO. As part of the formation of the Symbian OS in 2009 these three UIs were merged into a single OS which is now fully open source. Recently, though shipments of Symbian devices have increased, the operating system's worldwide market share has declined from over 50% to just over 40% from 2009 to 2010. Nokia handed the development of Symbian to Accenture, which will continue to support the OS until 2016.[39]

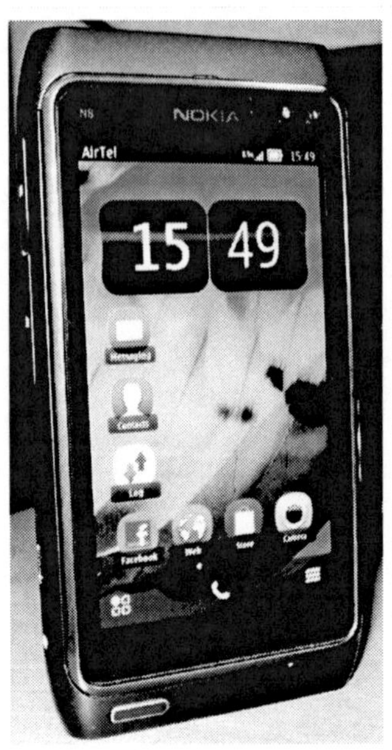

The Nokia N8 running Nokia Belle (Symbian OS 9.5.2)

bada from Samsung Electronics (closed source, proprietary)

This is a mobile operating system being developed by Samsung Electronics. Samsung claims that bada will rapidly replace its proprietary feature phone platform, converting feature phones to smartphones.The name 'bada' is derived from □ □ , the Korean word for ocean or sea. The first device to run bada is called 'Wave' and was unveiled to the public at Mobile World Congress 2010. The Wave is a fully touchscreen running the new mobile operating system. With the phone, Samsung also released an app store, called Samsung Apps, to the public. It has close to 3000[40] mobile applications.

Samsung has said that they don't see Bada as a smartphone operating system, but as an OS with a kernel configurable architecture, which allows the use of either a proprietary real-time operating system, or the Linux kernel. Though Samsung plans to install bada on many phones, the company still has a large lineup of Android phones.

Windows Phone from Microsoft (closed source, proprietary)

On February 15th, 2010, Microsoft unveiled its next-generation mobile OS, Windows Phone. The new mobile OS includes a completely new over-hauled UI inspired by Microsoft's "Metro Design Language". It includes full integration of Microsoft services such as Windows Live, Zune, Xbox Live and Bing, but also integrates with many other non-Microsoft services such as Facebook and Google accounts. The new software platform has received some positive reception from the technology press.[41][42][43]

Other software platforms

webOS from HP (certain parts open sourced)

webOS is a proprietary mobile operating system running on the Linux kernel, initially developed by Palm, which launched with the Palm Pre. After being acquired by HP, two phones (the Veer and the Pre 3) and a tablet (the TouchPad) running webOS were introduced in 2011. On August 18th, 2011, HP announced that webOS hardware is discontinued[44] but HP will continue to support and update webOS software and develop the webOS ecosystem.[45] HP plans to open-source webOS in the future.

GridOS from Fusion Garage[46]

GridOS was built using open source code from the Android kernel.[47] It is used as the operating system of the Grid 4 mobile phone and Grid 10 tablet.

Mer project (community-driven)

Mer is an open, mobile-optimised, core distribution aimed at device manufacturers; powered by Qt/QML and HTML5 – openly developed, inclusive, and meritocratically governed.

BlackBerry Tablet OS from QNX Software System/RIM[30] (closed source, proprietary)

BlackBerry Tablet OS is based on QNX. QNX is a commercial Unix-like real-time operating system, aimed primarily at the embedded systems market. The product was originally developed by Canadian company, QNX Software Systems, which was later acquired by BlackBerry-producer Research In Motion. This OS is primary for BlackBerry Playbook (tablet).

SHR (community driven)

SHR is a GNU/Linux based operating system for smartphones and similar mobile devices. It integrates various Free and Open Source Software projects into a versatile platform – flexible enough to run on a vast selection of mobile hardware such as the Openmoko Neo Freerunner, Nokia N900, Palm Pre and variants, T-Mobile G1, HTC HD2, iPhone 3Gs and more. The SHR build system is based on OpenEmbedded – well known from the Yocto project. For telephony, networking, etc. freesmartphone.org framework is used. On top of that an easy to use graphical interface centered around Enlightenment libraries is used to provide phone calls, messaging and pim. A growing amount of available applications offers SHR users with everything expected on a smartphone. But also numerous classical programs well known from other Linux distros can be made available easily.

SHR is 100% community driven and based on Free and Open Source Software. This empowers everybody to realize their innovations or add support for new hardware – without needing to ask some CEO or strategy manager first.[48]

MeeGo from non-profit organization The Linux Foundation (open source, GPL)

At the 2010 Mobile World Congress in Barcelona, Nokia and Intel both unveiled 'MeeGo' a brand new mobile operating system which would combine the best of Moblin and the best of Maemo to create a truly open-sourced experience for users across all devices. As of 2011, Nokia has announced that it will no longer be pursuing MeeGo and will instead adopt Windows Phone as its primary mobile OS. Nokia announced the Nokia N9 on June 21, 2011 at the Nokia Connection event[49] in Singapore. The phone is presumed to become available to the public in September 2011. LG announced its support for the platform.[50]

Linux based operating system (open source, GPL)[30]

Linux is strongest in China where it is used by Motorola, and in Japan, used by DoCoMo.[51] Rather than being an OS in its own right, Linux is used as a basis for a number of different operating systems developed by several vendors, including Android, GridOS, Boot to Gecko, LiMo, Maemo, MeeGo, Openmoko and Qt Extended, which are mostly incompatible.[52][53] PalmSource (now Access) is moving towards an interface running on Linux.[54] Another software platform based on Linux is being developed by Motorola, NEC, NTT

DoCoMo, Panasonic, Samsung, and Vodafone.[55]

Brew from Qualcomm

Brew is used by some mobile phone manufacturers and mobile networks, however most often the end-user does not know this since mobile phones running Brew most often lack any Brew branding. Brew runs in the background with the custom "skins" of the mobile phone manufacturer or operator on-top. Brew is used by Sprint Nextel, metroPCS, U.S. Cellular and Verizon in the US and by the Three network in much of Europe, the UK and Australia on many mobile phones produced especially for their network. Manufacturers such as LG CDMA, Huawei, INQ Mobile, Amoi, and Samsung Mobile amongst others use Brew in some of their mobile phones and it is featured in Three UK phones such as the 3 Skypephone, INQ1 and Huawei u7510 (3 Touch). Two of HTC's mobile phones use Brew's successor Brew MP.

LiMo 4 from LiMo Foundation

LiMo Foundation launched LiMo 4 on 14 february 2011, the latest release of the LiMo Platform. LiMo 4 delivers complete middleware and base application functionality, including a flexible user interface, extended widget libraries, 3D window effects, advanced multimedia, social networking and location based service frameworks, sensor frameworks, multi-tasking and multi-touch capabilities. In addition, support for scalable screen resolution and consistent APIs means that the platform can deliver a consistent user experience across a broad range of device types and form factors[56]

Historical software platforms

Maemo from Nokia (open source, GPL)

Maemo is a software platform developed by Nokia for smartphones and Internet tablets. It is based on the Debian operating system.

Maemo is mostly based on open source code, and has been developed by Maemo Devices within Nokia in collaboration with many open source projects such as the Linux kernel, Debian and GNOME.

Maemo is based on Debian GNU/Linux and draws much of its GUI, frameworks and libraries from the GNOME project. It uses the Matchbox window manager and the GTK-based Hildon as its GUI and application framework.

Windows Mobile from Microsoft[30][57] (closed source, proprietary)

The Windows CE operating system and Windows Mobile middleware are widely spread in Asia. The two improved variants of this operating system, Windows Mobile 6 Professional (for touch screen devices) and Windows Mobile 6 Standard, were unveiled in February 2007. It has been criticized for having a user interface which is not optimized for touch input by fingers; instead, it is more usable with a stylus. However, unlike iOS, it does support both touch screen and physical keyboard configurations.

Windows Mobile's market share has sharply declined in recent years to just 5% in Q2 of 2010.[58] Microsoft is phasing out the Windows Mobile OS to specialized markets and is instead focusing on its new operating system and software platform, Windows Phone.

Palm OS/Garnet OS from Access Co. (closed source, proprietary)

webOS was introduced by Palm in January 2009 as the successor to Palm OS with Web 2.0 technologies, open architecture, and multitasking capabilities.

DangerOS from Danger, Inc. (closed source, proprietary)

DangerOS was a largely Java-based operating system for the Danger Hiptop line of smartphones produced by Danger Incorporated from 2002 to 2010, also sold as the T-Mobile Sidekick. In 2008, Danger, Inc. was acquired by Microsoft.[59] Former Danger, Inc. employees were tasked to work on the Microsoft Kin line of phones, and in 2010 DangerOS was discontinued as a smartphone platform after the Kin phones were released.

In 2011 T-Mobile introduced a new smartphone using the Sidekick branding based on Google's Android platform, with no relation to the previous Danger Hiptop phones.[60]

Nucleus from Mentor Graphics (closed source)

Nucleus OS is a real-time operating system (RTOS) and toolset created by the Embedded Systems Division of Mentor Graphics for various central processing unit (CPU) platforms. The Nucleus RTOS is designed for embedded systems applications including consumer electronics, set-top boxes, cellular phones, and other portable and handheld devices. For limited memory systems Nucleus RTOS can be scaled down to a memory footprint as small as 13 KB for both code and data. Manufacturers such as Motorola, Samsung, LG, Siemens/Benq, Sagem, Pantech, and NEC use Nucleus in some of their mobile phones. The last devices released that used Nucleus OS are LG Encore (GT550), LG Prime (GS390), and Pantech Pursuit (P9020).

Upcoming software platforms

Firefox OS[61] from non-profit organization Mozilla Foundation (open source, GPL)

According to Ars Technica, "Mozilla says that B2G is motivated by a desire to demonstrate that the standards-based open Web has the potential to be a competitive alternative to the existing single-vendor application development stacks offered by the dominant mobile operating systems."[62]

BlackBerry 10 from RIM

BlackBerry 10 (previously BlackBerry BBX) the next generation platform for BlackBerry smartphones and tablets. In other words, there will be only one OS for both Blackberry smartphones and tablets going forward.[63]

Aliyun OS from Alibaba/AliCloud (cloud based)

AliCloud's operating system revolves around the idea of bringing cloud functionality to the mobile platform. According to the company, Aliyun will feature cloud-based e-mail, Web search, weather updates, and GPS navigation tools. In addition, the operating system will synchronize and store call data, text messages, and photos in the cloud for access across other devices, including PCs. Alibaba says it will offer customers 100GB of storage at launch. the operating system would allow users to access applications from the Web, rather than download apps to their devices.[64]

Tizen from non-profit organization The Linux Foundation (open source, GPL)

Tizen is an open source project hosted by the Linux Foundation, with Intel and Samsung leading its development steering group, and support from the LiMo Foundation. According to Intel, Tizen "combines the best of LiMo and MeeGo." HTML5 apps will be emphasized for the new OS, with the MeeGo project encouraging its members to transition to Tizen, stating that the "future belongs to HTML5-based applications, outside of a relatively small percentage of apps, and we are firmly convinced that our investment needs to shift toward HTML5." Tizen will be targeted at a variety of platforms such as handsets, tablets, smart TVs and in-vehicle entertainment. The initial release of Tizen is targeted for Q1 2012, with the first devices using the OS planned to reach the market in mid 2012.[65][66]

Windows 8 from Microsoft (closed source, proprietary)

Microsoft announced that Windows 8 will support tablet devices as well as PCs.

Windows RT from **Microsoft (closed source, proprietary)**

Microsoft announced Windows RT. an OS design for only tablets and can only run on arm. This version also resembles windows 8. Windows RT cannot run x86 programs. Apps can be only downloaded from the windows store. This version also has Microsoft office 2013 preinstalled on every Windows RT device.

Mobile OS comparison

Only the latest versions are shown in this table, even though old versions may still be marketed.

Legend

Yes	Tablet-only	Beta	Limited	3rd party	No

Feature	iOS	Android	webOS	Windows Mobile	Windows Phone	BlackBerry OS	Series40	Symbian	Maemo	MeeGo	
Company	Apple	Open Handset Alliance/Google	Palm, Inc (HP)	Microsoft	Microsoft	RIM	Nokia	Symbian Foundation	Nokia	Linux Foundation	
Current Version	5.1.1	4.1.1	2.2.4 (Phones) 3.0.5 (Tablet)	6.5.3	7.5	7.0.0	Developer Platform 2.0	10.1	5.0	1.1.2	
OS Family	Darwin	Linux	Linux	Windows CE 5.2	Windows CE 7 / Windows NT 8+[67]	Mobile OS	Proprietary OS	Mobile OS	Linux	Linux	Pro
Supported CPU Architecture	ARM	ARM, MIPS, Power Architecture, x86	ARM	ARM	ARM	ARM	?	ARM	ARM	ARM, x86	
Programmed in	C, C++, Objective-C	C, C++, Java	C	Many, C++, .NET, Lazarus	C,C++ 8+[68], .NET (Silverlight/XNA)	Java	?	C++	C/C++	C++	
License	Proprietary EULA except for open source components	Free and open source[31]	Free and open source except closed source modules	Proprietary	Proprietary	Proprietary	Proprietary	Eclipse Public License	Free and open source except closed source components[69]	Free and open source	F
Public issues list	No, But there is an unofficial tracker[70]	Yes[71]	No	No	Yes[72]	No	No	Not anymore[73]	Yes[74]	Yes[75]	
Package manager	iTunes	APK	App Catalog (Official) Preware (3rd party homebrew)	Windows Mobile Device Center/ActiveSync	Zune Software	BlackBerry Desktop Manager	Nokia Suite	Nokia Suite	dpkg+apt-get	rpm+yum+zypper	Sa
Fine grained storage usage	5+[76]	Yes[77]	?	?	No	?	?	?	?	?	
Per application network data usage auditing	3rd party software[78]	4+[79]	?	3rd party software[80]	3rd party apps on 8+ like Nokia Counters[81]	No[82]	No	3rd party software[83]	?	?	
Wireless system updates	5+[84]	Yes	Yes	No	8+[85]	Yes[86]	Yes	Yes[87]	Yes[88]	Yes	
Multi-user	No[89]	No[90], but coming on next release[91]	No[92]	No	No	No	No	No	No	No[93]	
Guest Mode	6+[94]	No,[95] but possible with 3rd party ROM MIUI[96]	No	No	No	No	No	No	No	No	

cation odifiable ions	Location access;[97] Notifications in 5+[98][99] Contacts, calendar and photos in 6+[100]	No,[101] only Background Data Usage on 4+,[79] more with 3rd party firmwares[102]	No[103]	?	No[104]	Yes[105]	?	?	?	?	?
ontrols	Yes[106]	3rd party software[107]	?	?	?	?	No	?	?	?	?
e location ng	Yes[108]	3rd party software[109]	?	?	Yes[110]	Yes	No	?	?	?	?
ce locking ta wipe	Yes[108]	Only Exchange 2.2+[111]	Yes[112]	Yes	Yes[110]	Yes	No	Yes	?	?	?
languages rt	Yes[113][114][115]	Limited[116]	Yes	Yes	Yes	Yes	Yes	Yes	Yes	Yes	Limited (Search is not diacritical mark insensitive)
ng spell er	Yes[117]	4+[118][119]	Yes	?	Yes[120]	Yes	Yes	Yes	Yes	?	No
em-wide ary	5+[121]	Built into keyboard application. Available on every device.[122]	?	?	?	Yes	?	?	?	?	No
ding text nents	5+[123]	?	Yes[124]	?	?	Yes	No	?	?	?	?
pport	Yes[125]	No,[126] but 3rd party apps available[127]	?	?	Yes[128]	?	?	?	?	?	?
features	Yes[129][130][131]	Yes	?	?	Yes	Yes	?	?	?	?	?
ultiple ications at	Yes	Yes	Yes	Yes	No[132]	Yes	No	Yes	Yes	?	Yes
ications earch	Bookmarks[133]	Calendar[134] (available since 3+ and HTC Sense)	?	?	Text Messages[135] nor Calendar,[136] nor Music nor Video[137]	?	?	?	?	?	?
fields of bjects	6+[138]	Only search Contacts name field but not any other field[139]	Yes	Yes	Contacts are only search by name[140]	Limited on most models[141] Search on notes body and contacts notes field was removed on OS6[142]	?	Yes	Yes	?	Only searches contacts by name
erver	Yes	3.1+[143] but only global, not per connection[144]	No[145]	Yes	Yes	Yes	Yes	Yes	Yes	?	Yes

On-device encryption	Yes[146] (3rd party software may attempt brute-force attacks on password[147])	3+[148]	No	Yes[149]	8+[150]	Yes[151] (3rd party software may attempt brute force attacks on password[147]	?	Yes	Yes	?
External storage encryption	External storage not supported	No[152]	External storage not supported	?	External storage not supported	Yes[153]	No	Yes	?	?
Cloud encrypted storage	No, data stored on iCloud is accessible by Apple[154]	No, data is accessible by Google and US Intelligence Agencies[155]	?	?	No, data is accessible by Microsoft and the US Government[156]	?	No	?	?	?
Privacy of synchronization[157]	When synchronizing locally and not using iCloud (that keeps data unencrypted)	Not possible since there is only Google synchronization (that keeps data unencrypted), possible with HTC Sense Desktop Sync	?	Yes	No, Contacts, Calendar and Mail can be kept private when using your own Exchange server instead of synchronizing with public cloud servers (that keeps data unencrypted)[158] but Photos and Office docs are uploaded to SkyDrive that cannot be disabled[159]	When using your own BES instead of BlackBerry cloud BES (that keeps data unencrypted)[160]	When synchronizing locally	When synchronizing locally and not synchronizing with Ovi (that keeps data unencrypted)	?	?
Sync to Cloud communication encryption	Yes[161]	2.3.4+[162]	?	?	7.10.7720.0+[163]	Yes[164]	?	?	3rd party software	?
Desktop Sync	Yes	No,[165] but available on HTC Sense[166]	3rd party software[167]	Yes	No[168][169]	Yes	Yes	Yes	Yes[170]	Yes
Local Full backup	Yes, using external computer[171]	No,[172] but possible with rooted devices and 3rd party software[173]	Limited official backup. Full backup via homebrew[174]	Yes	No[175]	Yes, using external computer[176] and memory card[177] (7+)	?	Yes, using external computer and memory card	?	?
Core Data missing Sync		Bookmarks[179] (before 4), SMSs and Settings[172]			Favorites,[180] Text Messages[181] and Settings[182]	?	?	?	?	?
cut, copy, and paste	Yes[183]	Yes[184]	Yes	Yes	Yes[185][186]	Yes	Yes	Yes	Yes	?
undo	Yes[188]	No[189]	Limited	Via Ctrl+Z using default keyboard	No	No	No	No	Yes	?
Visual voicemail	Yes	4+[190]	?	?	Yes[191]	Yes	Yes[192]	?	?	?
Call log duration	Yes[193]	Yes[194]	?	?	No[195]	Yes[196]	Yes	Yes	?	?
Multiple Mobile Phones per Contact	Yes	Yes	?	?	No[197]	?	Yes	?	?	?

ups	3rd party software[198]	4+[199] or 3rd party software[200]	?	?	7.5+[201]	Yes	Yes	Yes	?	?	Yes
linking	Yes	Available in stock Android, but not all devices enable it.[202]	Yes	Yes, but not in browser	Yes	Yes	Yes	Yes	Yes	?	Yes, but not in browser
ent	No	Available on any device that supports the hardware. Google Wallet for NFC payments available in Play Store.[203]	No	No	8+[204]	No	No	No	No	No	No
o maps	Yes[205]	In Contacts, but not in Calendar[206] nor in Gmail[207]	?	?	Yes[208]	?	?	In Contacts and Calendar	?	?	?
lendar	Yes[209]	No	?	?	?	?	?	?	?	?	?
pport	Yes[210]	No,[211] but 3rd party app available[212]	?	3rd party apps available[213]	No[214]	Yes[215]	Yes[216]	?	?	?	?
eb gine	Webkit	Webkit	Webkit	Trident	Trident	Webkit	Webkit	Webkit	Gecko	Webkit	Webkit
ect ection ence to	Yes	Yes	Yes	Yes	Yes	No, routes through BlackBerry Internet Service. Direct browsing supported over WiFi	Yes	Yes	Yes	Yes	Yes
eps n on rash	Yes	4.1+Google Chrome[217] but not on main browser[218]	No	No	Yes	?	No	?	Yes	?	No
arch ons	Bing, Google, Yahoo! Search	Many[219]	Many[220]	?	Bing, Google[221]	Google, Bing, Yahoo, Wikipedia, Dictionary.com, User Defined	Google	Google	?	?	?
n Page	4.2+[222]	1.5+[223]	?	?	Removed in Mango update[224]	Yes[225]	?	Yes	?	?	?
l Save	Yes[226]	Yes	?	?	Yes[227]	Yes	?	Yes	?	?	?
l Save	Yes[228]	Yes	?	?	Yes[229]	Yes	?	Yes	?	?	Yes
ive eo	No	Yes, but only links, not embedded media	Yes	?	No	Yes (embedded media not supported)	?	?	?	?	Yes
Page	6+ Offline Reading List[230] or 3rd party software[231]	4+[232]	No	?	No	Yes[233]	?	Yes	?	?	?
ny File	No, 3rd party software[234]	No,[235] 3rd party software[236]	?	?	No, 3rd party software[237]	?	?	?	?	?	?

Browser Text Reflow	Only on pages available on reader mode on 5+:[238][239] 3rd party browsers on prior versions[240]	Yes[241]	No[242]	?	No[241]	Yes[243]	?	No	?	?
Browser Reader Mode	Not available on every page[238][239]	?	?	?	?	Beta PlayBook OS 2.0[244]	?	?	?	?
Browser file upload	6+[230] or 3rd party software[245]	2.2+,[246] but crashes on big files[247]	?	?	Yes	Yes	Yes	Yes	?	?
Browser form navigation	Previous, Next, AutoFill, and Done buttons[248]	Next button[249]	?	?	No[250]	?	?	?	?	?
Browser Private Browsing mode	5+[251]	3.0+[252] or with rooted devices and 3rd party firmware[253]	?	?	3rd party software[254]	?	?	?	?	?
Offline web apps	Yes[255][256][257]	Yes[258]	?	?	?	?	?	?	?	?
Official Application Store	App Store	Google Play	App Catalog	Windows Marketplace for Mobile	Windows Phone Marketplace	App World	Nokia store	Symbian Horizon, Nokia store	maemo.org, Nokia store	?
Non-discriminatory Stores	Apple discriminates based on country[259] and own Apple policies[260]	?	Discriminates by country.[261] Can be circumvented via homebrew[262]	?	Discriminates by country[263]	3rd party software	?	?	?	?
Unified Inbox	Yes	No,[264] but 3rd party apps available[265]	Yes	Yes	7.10.7720.0+[266]	Yes	No	Yes	?	?
Email Sync protocols supported	POP3, IMAP, MAPI	POP3, IMAP, MAPI	POP3, IMAP, MAPI	POP3, IMAP, MAPI	POP3, IMAP, MAPI	BES, BIS, Push e-mail	POP3, IMAP, Exchange	POP3, IMAP, Exchange	POP3, IMAP, Exchange	POP3, IMAP
Non-intrusive Notifications	5+[267]	Yes[268]	Yes[269]	?	Yes,[270] except Calendar Reminders[271]	6+[272]	Yes	Symbian Belle+[273]	Yes[274]	Yes[275]
Notification Center	5+[267]	Yes	Yes[276]	?	No[277]	6+[278]	Yes	Yes	Yes	Yes
Push Notifications	Yes (Apple Push Notification Service)	Yes[280]	Yes	?	Yes[281]	Yes	Yes	Yes	?	?
End-to-end Encrypted Push Notifications	Not possible, Apple must have access to notifications cleartext[283]	Possible, notifications are handled by the application that can decrypt it[280]	?	?	Possible, raw notifications are handled by the application that can decrypt it[284]	?	?	?	?	?
Screen orientation lock	Yes[228]	Yes[285]	?	?	No[286]	3rd party software	No	Yes	?	?
Custom Home and Lock Screen wallpaper	4+[228]	Yes	?	?	Yes	Yes	Yes	Yes	?	?
Separate volume for Ringtone and Media	Yes	Yes	?	?	No[287]	?	Yes	?	?	?

ngtones	Yes[288]	Yes	Yes	Yes	7.10.7720.0+[289]	Yes	Yes	Yes	3rd party software[290]	?	Yes
IS/MMS s	5+[291]	Yes[292]	Via homebrew[293]	?	No[294]	Yes[295]	Yes	Yes	?	?	?
rate Alert ns	5+[296][297]	No[298]	?	?	?	Yes[295]	No	?	?	?	?
r-based essaging	5+ (iMessage)[299]	Google Talk	AIM, Google Talk, Yahoo	?	Windows Live Messenger, Facebook Chat[300]	BBM	Many (Google Talk, Yahoo)	Many	?	?	?
Voice nds	Yes[301]	No[302]	2.1+	Microsoft Voice Command	Limited, Playback control not available[303]	Yes	?	Yes	?	Yes	?
gnition	5+ (Siri on iPhone 4S)[304]	Yes[305]	2.1+	Yes	Yes[306][307]	Yes	No	Yes	?	Yes	?
Voice tion	No[308]	4.1+[309]	No	No	No	No	No	No	No	No	{{No}}
h Voice tion	Yes[115]	Yes	?	?	Yes[128]	?	No	?	?	?	?
corder	Limited (Not voice controlled)	Very limited (Doesn't work in background[310] and not voice controlled)	3rd party software	?	3rd party software; very limited (Works under locked screen but doesn't work in background and not voice controlled)	Very limited (Doesn't work in background and not voice controlled)	Yes	Yes	3rd party software[311]	?	Neutral
order	Prevented by OS restrictions	Prevented by OS restrictions,[312] but possible with 3rd party firmware[313]	No	Yes[314]	No	No[315]	Yes	3rd party software[316]	3rd party software[311]	?	?
to-focus	Yes[228]	4+[317]	?	?	Yes[318]	?	?	?	?	Yes[319]	?
cture info xif data)	Only location[320]	Only date and location; more with 3rd party software[321]	Yes	?	No	?	?	Not full Exif	?	?	?
picture ation	5+[322]	No[323]	3.0+	?	No[324]	Yes, photos can be renamed and searched	?	Yes, albums and tags[325]	?	?	yes(by date or manual)
picture nd search	No	No[326]	?	?	No	?	?	No, Allows setting a description, but search will not use it	?	?	?
tter sound	In silent mode[327]	In silent mode;[328]	In silent mode	?	7.10.7720.0+[329]	?	?	In silent mode	?	?	Yes(in manual & in Silent)
crop	5+[330]	Yes	3rd party software	?	3rd party software[331]	?	?	Yes	Yes	?	Yes

Photo rotation	5+[330]	Yes	?	?	3rd party software[332]	?	Yes	Yes	?	?	
Photo red-eye reduction	5+[330]	4+[333]	?	?	3rd party software[334]	?	?	Symbian Anna+[335]	?	?	
HDR photos option	Yes[336]	No,[337] 3rd party software available[338]	?	?	?	?	No	?	?	?	
Photo/video import from memory cards	With Camera Connection Kit[339]	No	?	?	?	Yes	Yes	Yes	?	?	
Video trim	Yes	Yes3.0+[340]	Yes	?	No	?	?	Yes[341]	?	?	
Sound trim	Yes	No,[342] but 3rd party software available[343]	No	?	No sound recorder	?	?	?	?	?	
Tethering	Bluetooth, USB (carrier dependent), Personal Hotspot (Wi-Fi Tethering) (carrier dependent, iPhone 4 & 4S since iOS 4.2.5/4.3, or with 3rd party software and "jailbreak")	Mobile Wi-Fi Hotspot, USB, Bluetooth	Mobile Wi-Fi Hotspot (officially AT&T and Verizon Wireless only). USB, Bluetooth, and Mobile Wi-Fi Hotspot via homebrew[344]	USB, Bluetooth, Mobile Wi-Fi Hotspot (with 3rd party software)	Internet Sharing (Wi-Fi Hotspot)[345]	USB, Bluetooth, Mobile Wi-Fi Hotspot	USB, Bluetooth	USB, Bluetooth, Mobile Wi-Fi Hotspot (with 3rd party software)	microUSB, Bluetooth, Mobile Wi-Fi Hotspot	?	B M
USB On-The-Go	No[346]	3.1+[347][348]	No	?	No	?	Yes	Yes[349]	?	?	
Interchangeable external memory cards	Only for photo/video import with an optional accessory	Yes[350]	No	Yes	8+[351]	Yes[352]	Yes	Yes	Yes	Yes	
Text/Document Support	Microsoft Office,[354][355][356] iWork, PDF, Images, TXT/RTF, VCF	Microsoft Office 2003/2007, PDF, Images, TXT/RTF	Microsoft Office, PDF, TXT/RTF	Microsoft Office Mobile, PDF, TXT/RTF	Microsoft Office Mobile, PDF, TXT/RTF	Microsoft Office, PDF, OpenDocument	with free 3rd party software[357]	Microsoft Office Mobile, PDF, djvu	text files, PDF, HTML, Multiple office formats with free 3rd party software	?	Re HT of
Printer support	Yes (AirPrint)[358]	No built-in function,[359] but have apps use Google Cloud Print available for 2.0+[360]	Yes[361]	?	No,[362] but 3rd party software available	?	No	?	?	?	

yback	AAC (8 to 320 kbit/s), Protected AAC (from iTunes Store), HE-AAC, MP3 (8 to 320 kbit/s), MP3 VBR, Apple Lossless, AIFF, WAV	AAC LC/LTP 3GPP, HE-AACv1 (AAC+) (before Android2.1),[363] HE-AACv2 (enhanced AAC+)(before Android2.1),[363] AMR-NB, AMR-WB, MP3 (Mono/Stereo 8–320 kbit/s constant or variable bit-rate), MIDI (Type 0 and 1, DLS versions 1 and 2), Ogg Vorbis, PCM/WAVE (8- and 16-bit linear PCM, rates up to limit of hardware), WAVE[364]	3G2, A-LAW, AAC, AAC LC, AAC LTP, AAC+, AMR-NB, AMR-WB, AWB, HE-AAC v1, HE-AAC v2, LPCM, M4A, MIDI Tones (poly 64), Mobile XMF, MP3, MP4, MU-LAW, NRT, SP-MIDI, WAV, WMA 10, WMA 10 Pro, WMA 9, X-Tone	MP3, AAC, AAC+, AMR, QCELP, WAV		MP3, AAC, AAC+, eAAC+, WAV, WMA pro, AMR-NB, MIDI	MP3, WAVE, WMA, AAC+, MIDI, AMR, eAAC+, FLAC, OGG	All	All (some require optional debian packages)	?	MP3, AAC, WMA, M4A, XMF, 3GA, MMF, MIDI, WAV, AMR[365]
yback	H.264 AVC, MPEG-4, M-JPEG	H.263, H.264 AVC, MPEG-4 SP, DivX, XviD, VP8[364]	MPEG-4, H.263, H.264		H.263, H.264, WMV, MPEG4, MPEG4 @ HD 720p 30fps, DivX, XviD	MP4, WMV, H.263, H.264, DivX, WMV, XviD, 3gp	3GPP formats (H.263), ASF, AVI, H.264/AVC, MPEG-4, VC-1, WMV	H.263, H.264, WMV, MPEG4, MPEG4 @ HD 720p 30fps MKV, DivX, XviD	All (some require optional debian packages)	?	WMV, ASF, MP4, 3GP, AVI[365]
eo out	Up to 720p via HDMI[366] or VGA,[367] 576p/480p via component[368] or composite;[369] 1080p on iPad 2 or iPhone 4S in 5+	1080p on select devices	No	?	No[370]	No	No	Nokia AV Out (PAL/NTSC), HDMI	Nokia AV Out (PAL/NTSC)	?	usb2hdmi & AV Out (PAL/NTSC)
deo/audio o set top speakers	AirPlay[371]	No,[372] DLNA available on selected devices	?	?	No,[373] Play To (DLNA) available on some devices[374]	?	No	?	?	?	DLNA[375]
layer playlist on	Yes[228]	Yes[376]	?	?	Yes[377]	Yes	Yes	?	?	?	?
er Video ing	Yes	Yes	No	?	7.10.7720.0+[378]	?	?	?	?	?	?
er Audio ing	Yes	Yes	No	?	No	?	?	?	?	?	?
ver Fine ing	Yes[379]	3rd party software[380][381]	No	?	No	?	?	?	?	?	?
er Double aying	Only Audio Podcasts[382]	3rd party software[383][384]	?	?	No	?	?	?	?	?	?

Turn-by-turn navigation	6+[385]	2+[386]	Carrier software, 3rd party software	3rd party software	8+[387], limited in WP7[388], but needs a tap at each turn[389] and maps outside USA are outdated,[390] or 3rd party software[391]	3rd party software	?	free global Nokia Maps and 3rd party software	Nokia Maps and 3rd party software	?	S
Offline maps	3rd party software[392]	Google Maps (will not work with limited phone coverage[393] as it only downloads a 10-mile radius[394]) or 3rd party software	Carrier software, 3rd party software	3rd party software	8+[395] 3rd party software[396]	3rd party software	free global Nokia Maps and 3rd party software	free global Nokia Maps and 3rd party software	Nokia Maps and 3rd party software	?	S
Alternative routes in maps	5+[397]	Yes[398]	?	?	3rd party software	?	?	?	?	?	
Multitasking	Very Limited, Apple-approved functionality only. OS 4.0+[399]	Yes	Yes	Yes	8+[400]	Yes	Limited	Yes	Yes	Yes	
Record video with voice while calling[401]	No	Yes (Not available on some devices)	?	?	Yes	?	No	No	?	?	
Desktop interactive widgets	No	Yes	Via homebrew	Yes	Yes (through "live tiles")[402]	No	Yes	Yes	Yes	Yes	
Lock screen widgets	Media player; 5+: Notifications,[267] voicemail, camera[403]	Notification and Camera	?	?	?	?	No	?	?	?	M Not
Notification view widgets	5+: Stocks and weather;[267] 3rd party software with "jailbreak"[404][405]	?	3.0+[406]	3rd party firmware	No (there is no notification view)	?	No	Yes	?	?	
Bluetooth keyboard	Yes[407]	3.1+;[408] previous versions via 3rd party software[409]	2.0+	Yes	No[410]	Yes	No	Yes	Yes, with plugins	?	
USB keyboard	With Camera Connection Kit[339]	3.1+[408]	?	?	No[411]	?	No	Yes	?	?	
Direct file transfer	Bluetooth using 3rd party software on jailbroken devices[412]	Bluetooth,[413] Wi-Fi Direct 4+[79] and selected devices[414][415] and NFC (small data) 4+[79]	No	Bluetooth	No[416]	Bluetooth	Bluetooth	Bluetooth	Bluetooth	Bluetooth	Yes 2
Voice over IP	FaceTime or 3rd party software (like Skype)	Yes (SIP)[418] or 3rd party software (like Skype)	Yes[419]	3rd party software (like fring)	Fully integrated in 8+[420]	3rd party software[421]	Yes (SIP)[422]	Yes (SIP) or 3rd party software (like Skype)	Yes (SIP)[423]	?	
SSH	Yes[425][426]	Yes	Via homebrew	No	3rd party software	Yes	No	?	Yes	Yes	

	Yes[427]	Yes[428]	3rd party software[429]	Yes[430]	No[431]	No Only through BlackBerry Enterprise Server	?	Yes[432]	No[433]	No[434]	?
N	No[435]	No, but possible with 3rd party firmware[436]	Yes[437]	3rd party software[438]	No	No	?	No	Yes[439]	Yes[440]	?
AP	Yes[441]	Yes[442]	Yes[443]	?	Yes[444]	?	?	?	?	?	?
ered tions	No[445]	Yes	?	?	Yes	?	Yes	?	?	?	?
Buffer	?	Yes[446]	Via homebrew[447]	?	?	?	?	?	Yes	Yes	?
t	Yes[228][448]	4+ also available on 3.7 or earlier with Cyanogen Mod and on certain devices E.G Samsung Galaxy S II [449]	Yes	Yes[450]	No,[451] but possible through homebrew or SDK.[452][453]	3rd party software	?	Yes[454]	Yes	?	Yes
t	No, external hardware required[455]	No, root required and 3rd party app[456]	No	No	No	No	?	No[457]	No	No	No
ed GUI	Yes	3+[458]	Partially (in Enyo apps)	No	Yes[459]	Yes, OS 7.0+	?	Yes	Yes	?	?
K s)	Mac OS X using iOS SDK	Linux, Mac OS X and Windows[460]	Linux, Mac OS X and Windows[461]	Windows[462]	Windows[463]	Windows, Mac OS X[464]	Nokia S40 SDKs[465]	Windows using Symbian SDK[466] or Linux, Mac OS X and Windows using Nokia Qt SDK[467]	GNU/Linux[468]	GNU/Linux and Windows[469]	Windows[470]
evelop	Free ($99/year to distribute on App Store[471])	Free ($25 once to offer it on the Google Play[472])	Free[473]	Free	Free ($99/year to offer it on the Windows Phone Marketplace[474])	Free SDK, Free Signing Keys, Free Publishing in App World	Free (Nokia Store)	Free (1€ once to offer it in the Ovi Store[475])	Free	Free	Free
cate to ntry	Yes	Yes	Yes	Yes	No[476]	Yes	Yes	Yes	Yes	Yes	Yes
	iOS	**Android**	**webOS**	**Windows Mobile**	**Windows Phone**	**BlackBerry OS**	**Series40**	**Symbian**	**Maemo**	**MeeGo**	**bada**

References

[1] Jobs, Steve (2007-01-19). *Macworld San Francisco 2007 Keynote Address* (http://www.apple.com/quicktime/qtv/mwsf07/). San Francisco: Apple, Inc.. .

[2] Helft, Miguel (2007-11-05). "Google Enters the Wireless World" (http://www.nytimes.com/2007/11/05/technology/05cnd-gphone.html?ex=1352005200&en=d7a169e184415788&ei=5088&partner=rssnyt&emc=rss). New York Times. . Retrieved 2011-09-07.

[3] "64 million smart phones shipped worldwide in 2006" (http://www.canalys.com/newsroom/64-million-smart-phones-shipped-worldwide-2006). Canalys, Inc. . Retrieved 2012-01-13.

[4] "Gartner Smart Phone Marketshare 2011 Q3" (http://www.gartner.com/it/page.jsp?id=1848514). Gartner, Inc. . Retrieved 2012-05-26.

[5] "Nokia has sold over 1.5 billion Series 40 phones" (http://www.esphoneblog.com/2012/01/25/nokia-has-sold-over-1-5-billion-series-40-phones/). . Retrieved 2012-01-25.

[6] 2010 Q4 contains insignificant part of Windows Phone devices

[7] not including low cost devices

[8] including low cost devices

[9] 2011 Q2 and the more contains insignificant part of Windows Mobile devices

[10] "Gartner Smart Phone Marketshare 2012 Q2" (http://www.gartner.com/it/page.jsp?id=2120015). Gartner, Inc. . Retrieved 2012-08-14.

[11] "Gartner Smart Phone Marketshare 2012 Q1" (http://www.gartner.com/it/page.jsp?id=2017015). Gartner, Inc. . Retrieved 2012-05-26.

[12] "Gartner Smart Phone Marketshare 2011 Q4" (http://www.gartner.com/it/page.jsp?id=1924314). Gartner, Inc. . Retrieved 2012-05-26.

[13] "Gartner Smart Phone Marketshare 2011 Q2" (http://www.gartner.com/it/page.jsp?id=1764714). Gartner, Inc. . Retrieved 2012-05-26.

[14] "Gartner Smart Phone Marketshare 2011 Q1" (http://www.gartner.com/it/page.jsp?id=1689814). Gartner, Inc. . Retrieved 2012-05-26.

[15] "Gartner Smart Phone Marketshare 2010 Q4" (http://www.gartner.com/it/page.jsp?id=1543014). Gartner, Inc. . Retrieved 2012-05-26.

[16] "Gartner Smart Phone Marketshare 2010 Q3" (http://www.gartner.com/it/page.jsp?id=1466313). Gartner, Inc. . Retrieved 2012-05-26.

[17] "Gartner Smart Phone Marketshare 2010 Q2" (http://www.gartner.com/it/page.jsp?id=1421013). Gartner, Inc. . Retrieved 2012-05-26.

[18] "Gartner Smart Phone Marketshare 2010 Q1" (http://www.gartner.com/it/page.jsp?id=1372013). Gartner, Inc. . Retrieved 2012-05-26.

[19] "Gartner Smart Phone Marketshare 2008 Q4" (http://www.gartner.com/it/page.jsp?id=910112). Gartner, Inc. . Retrieved 2012-05-26.

[20] "Gartner Smart Phone Marketshare 2008 Q3" (http://www.gartner.com/it/page.jsp?id=827912). Gartner, Inc. . Retrieved 2012-05-26.

[21] "Gartner Smart Phone Marketshare 2008 Q2" (http://www.gartner.com/it/page.jsp?id=754112). Gartner, Inc. . Retrieved 2012-05-26.

[22] Android- and iOS-Powered Smartphones Expand Their Share of the Market in the First Quarter, According to IDC – prUS23503312 (http://www.idc.com/getdoc.jsp?containerId=prUS23503312). Idc.com (2012-05-24). Retrieved on 2012-07-03.

[23] including low cost devices

[24] not including low cost devices

[25] including Bada OS

[26] http://www.idc.com/getdoc.jsp?containerId=prUS23638712

[27] http://www.idc.com/getdoc.jsp?containerId=prUS23638712

[28] Android Expected to Reach Its Peak This Year as Mobile Phone Shipments Slow, According to IDC | SYS-CON MEDIA (http://www.sys-con.com/node/2291309). Sys-con.com (2012-06-06). Retrieved on 2012-07-03.

[29] "Mobile Data Usage" (http://www.netmarketshare.com/operating-system-market-share.aspx?qprid=8&qpcustomd=1). . Retrieved 2012-07-01.

[30] "Gartner Says Worldwide Mobile Phone Sales Grew 35 Percent in Third Quarter 2010; Smartphone Sales Increased 96 Percent" (http://www.gartner.com/it/page.jsp?id=1466313). Gartner, Inc. 2010-11-10. Table 2. . Retrieved 2011-02-21.

[31] ICS is coming to AOSP – Android Building (http://groups.google.com/group/android-building/msg/c0e01b4619a1455a). Groups.google.com (2011-11-14). Retrieved on 2012-07-03.

[32] "Alliance Members" (http://www.openhandsetalliance.com/oha_members.html). Open Handset Alliance. . Retrieved 2011-09-07.

[33] "Developers Praise Android at Google I/O | PCWorld Business Center" (http://www.pcworld.com/businesscenter/article/146450/developers_praise_android_at_google_io.html?tk=rl_noinform). PCWorld.com. 2008-05-29. . Retrieved 2011-09-07.

[34] Android takes 52.5% of smartphone market (http://econsultancy.com/uk/blog/8279-android-doubles-market-share-to-take-52-5-of-smartphone-market). Econsultancy. Retrieved on 2012-07-03.

[35] Geller, Jonathan S. (2011-08-08). "RIM's first QNX phone revealed: BlackBerry Colt to launch in Q1 2012" (http://www.bgr.com/2011/08/08/rims-first-qnx-phone-revealed-blackberry-colt-to-launch-in-q1-2012/). BGR.com. . Retrieved 2011-09-07.

[36] "Nokia S40 platform" (http://www.developer.nokia.com/Develop/Series_40/Platform/). nokia.com. . Retrieved 2012-08-02.

[37] Tony Dennis (2006-10-18). "Clifford says Symbian will conquer US" (http://uk.theinquirer.net/?article=35179). theinquirer.net. . Retrieved 2011-09-07.

[38] Symbian Facts (http://www.symbian.com/about/fastfacts/fastfacts.html)

[39] "Nokia confirms Accenture to support Symbian until 2016 – 22 Jun 2011 – Computing News" (http://www.computing.co.uk/ctg/news/2080712/nokia-confirms-accenture-support-symbian-2016). Computing.co.uk. 2011-06-22. . Retrieved 2011-09-07.

[40] "Samsung s8530 Wave II review" (http://www.gsmarena.com/samsung_s8530_wave_ii-review-538p7.php). gsmarena.com. 2010-11-19. . Retrieved 2011-09-07.

[41] "Windows Phone 7 Series is official, and Microsoft is playing to win" (http://www.engadget.com/2010/02/15/windows-phone-7-series-is-official-and-microsoft-is-playing-to/). Engadget. 2010-02-15. . Retrieved 2011-09-07.

[42] Matt Buchanan (2010-02-15). "Windows Phone 7 Series: Everything Is Different Now" (http://gizmodo.com/5471805/windows-phone-7-series-everything-is-different-now). Gizmodo.com. . Retrieved 2011-09-07.

[43] Devin Coldewey (2010-02-15). "Windows Phone 7 Series: Our Take | TechCrunch" (http://www.crunchgear.com/2010/02/15/windows-phone-7-series-our-take/). Crunchgear.com. . Retrieved 2011-09-07.

[44] "HP Confirms Discussions with Autonomy Corporation plc Regarding Possible Business Combination; Makes Other Announcements" (http://www.hp.com/hpinfo/newsroom/press/2011/110818b.html?mtxs=rss-corp-news). HP. 2010-08-18. . Retrieved 2011-09-13.

[45] "The next chapter for webOS" (http://developer.palm.com/blog/2011/08/the-next-chapter-for-webos/). HP webOS Developer Blog. 2010-08-19. . Retrieved 2011-09-13.

[46] "fusiongarage" (http://www.fusiongarage.com). fusiongarage. . Retrieved 2011-09-07.

[47] "About GridOS" (http://www.fusiongarage.com/grid-os/about). fusiongarage. . Retrieved 2011-09-07.

[48] "SHR" (http://shr-project.org/trac). SHR-project.org. . Retrieved 2011-09-07.

[49] "Introducing the Nokia N9: all it takes is a swipe! | Nokia Conversations – The official Nokia Blog" (http://conversations.nokia.com/2011/06/21/introducing-the-nokia-n9-all-it-takes-is-a-swipe/). Nokia. 2011-06-21. . Retrieved 2011-09-07.

[50] "MeeGo Not Dead Yet as LG Continues the Charge — Mobile Technology News" (http://gigaom.com/mobile/meego-not-dead-yet-as-lg-continues-the-charge/). Gigaom.com. 2011-04-29. . Retrieved 2011-09-07.

[51] "Microsoft Excluded from DoCoMo's ecosystem" (http://www.theregister.co.uk/2004/11/26/microsoft_excluded_from_docomo/). The Register. 2004-11-26. . Retrieved 2011-09-07.

[52] "Incompatibility in Mobile Linux at "OS News"" (http://www.osnews.com/story.php/15040/Editorial-The-Chaos-of-Incompatibility-in-Mobile-Linux). OS News. 2006-06-29. . Retrieved 2011-09-07.

[53] "Mobile Platforms: Linux – a land of misconceptions" (http://searchmobilecomputing.techtarget.com/originalContent/0,289142,sid40_gci1218922,00.html). searchmobilecomputing.techtarget.com. 2006-09-27. . Retrieved 2011-09-07.

[54] "Can Linux save the Palm OS?" (http://news.com.com/Can+Linux+save+the+Palm+OS/2008-1045_3-6110042.html). CNET News. 2006-08-28. . Retrieved 2011-09-07.

[55] "Mobile Industry Leaders to Create World's First Globally Adopted Open Mobile Linux Platform" (http://mediacenter.motorola.com/Press-Releases/Mobile-Industry-Leaders-to-Create-World-s-First-Globally-Adopted-Open-Mobile-Linux-Platform-1e3c.aspx). *Press Release*. Motorola. 2007-01-25. . Retrieved 2011-09-07.

[56] LiMo 4 (http://www.limofoundation.org/en/Press-Releases/limo-foundation-unveils-limo-4.html)

[57] "CEO Ballmer Reportedly Says Microsoft 'Screwed Up' with Windows Mobile" (http://www.eweek.com/c/a/Windows/Microsoft-CEO-Steve-Ballmer-Says-Company-Screwed-Up-Windows-Mobile-241614/). *eWeek*. 28 September 2009. .

[58] "iPhone drops to 24% smartphone share, Android jumps to 17%" (http://www.appleinsider.com/articles/10/09/16/iphone_drops_to_23_8_smartphone_market_share_android_jumps_to_17.html). AppleInsider. 2010-09-16. . Retrieved 2011-09-07.

[59] "How Much Did Microsoft Pay For Danger?" (http://gigaom.com/2008/02/12/how-much-did-microsoft-pay-for-danger-find-out-here/). Gigacom. Feb. 12, 2008. . Retrieved October 19, 2011.

[60] Segan, Sascha (2011-01-20). "T-Mobile Confirms Sidekick 4G, Samsung Galaxy S 4G" (http://www.pcmag.com/article2/0,2817,2376019,00.asp). Pcmag.com. . Retrieved 2011-06-01.

[61] "B2G – MozillaWiki" (https://wiki.mozilla.org/B2G). mozilla.org. 2011-08-24. . Retrieved 2011-09-07.

[62] Paul, Ryan (2011-07-25). "Mozilla eyes mobile OS landscape with new Boot to Gecko project" (http://arstechnica.com/open-source/news/2011/07/mozilla-eyes-mobile-os-landscape-with-new-boot-to-gecko-project.ars). Arstechnica.com. . Retrieved 2011-09-07.

[63] BlackBerry BBX (http://blackberryrocks.com/2011/10/18/rim-unveils-blackberry-bbx-combines-blackberry-qnx-next-generation-platform/). Blackberryrocks.com (2011-10-20). Retrieved on 2012-07-03.

[64] Reisinger, Don (2011-07-28). "Alibaba OS-powered handset launching this month | The Digital Home" (http://news.cnet.com/8301-13506_3-20084745-17/alibaba-os-powered-handset-launching-this-month/). CNET News. . Retrieved 2011-09-07.

[65] Welcome to Tizen! (https://www.tizen.org/blogs/dawnfoster/2011/welcome-tizen). Tizen.org (2011-09-27). Retrieved on 2012-07-03.

[66] Ricker, Thomas. (2011-09-28) MeeGo is dead: Meet Tizen, another new open source OS based on Linux (http://thisismynext.com/2011/09/28/meego-dead-meet-tizen-open-source-os-based-linux/). Thisismynext.com. Retrieved on 2012-07-03.

[67] Windows Phone 8: the new features - 12. Shared Windows NT kernel, code with Windows 8 (http://www.phonearena.com/news/Windows-Phone-8-the-new-features_id31466#12-Shared-Windows-NT-kernel,-code-with-Windows-8)

[68] Windows Phone 8 adds DirectX, native development, easier app porting | The Verge (http://www.theverge.com/2012/6/20/3095150/microsoft-windows-phone-8-common-platform-directx)

[69] Open development/Why the closed packages – maemo.org wiki (http://wiki.maemo.org/Why_the_closed_packages). Wiki.maemo.org (2011-02-25). Retrieved on 2012-07-03.

[70] "PleaseFixTheiPhone" (http://pleasefixtheiphone.com/). PleaseFixTheiPhone. . Retrieved 2011-09-07.

[71] "Android Issues Tracker" (http://code.google.com/p/android/issues/list). Google.com. . Retrieved 2011-09-07.

[72] Have a new Windows Phone feature idea? Try Suggestion Box (http://windowsteamblog.com/windows_phone/b/windowsphone/archive/2011/09/28/have-a-new-windows-phone-feature-idea-try-suggestion-box.aspx)

[73] Symbian Foundation is completing its transition to a licensing body (http://blog.symbian.org/2010/12/17/symbian-foundation-is-completing-its-transition-to-a-licensing-body/)

[74] "Maemo Issues Tracker" (https://bugs.maemo.org/). Maemo.org. . Retrieved 2011-09-07.

[75] "MeeGo Issues Tracker" (http://meego.com/community/bug-tracking). MeeGo.com. . Retrieved 2011-09-07.
[76] How To See How Much Space Is Being Used By Each of Your Apps in iOS 5 — iPad Insight (http://ipadinsight.com/ipad-tips-tricks/how-to-see-how-much-space-is-being-used-by-each-of-your-apps-in-ios-5). Ipadinsight.com (2011-06-13). Retrieved on 2012-07-03.
[77] Android 4.0 Internal Memory Storage and Apps Management (http://www.androidtapp.com/whats-new-in-ice-cream-sandwich-android-4-0/android-4-0-internal-memory-storage-and-apps-management/). Android Tapp. Retrieved on 2012-07-03.
[78] App Store – My Data Manager (http://itunes.apple.com/us/app/my-data-manager/id477865384). Itunes.apple.com (2012-05-18). Retrieved on 2012-07-03.
[79] Android 4.0 Platform Highlights | Android Developers (http://developer.android.com/sdk/android-4.0-highlights.html). Developer.android.com. Retrieved on 2012-07-03.
[80] SPB Wireless Monitor » Pocket PC Software » SPB Software (http://spb.com/pocketpc-software/wirelessmonitor/). Spb.com (2010-01-26). Retrieved on 2012-07-03.
[81] Nokia announces a suite of new Windows Phone apps (http://blog.worldgsmphones.com/2012/06/23/nokia-announces-a-suite-of-new-windows-phone-apps/)
[82] Application Data Usage Monitoring – BlackBerry Applications (http://forum.shaarpmind.com/showthread.php/219924-Application-Data-Usage-Monitoring-BlackBerry-Applications). Forum.shaarpmind.com (2012-01-19). Retrieved on 2012-07-03.
[83] Full Data Traffic Control for Symbian with SPB Wireless Monitor » Press Releases » SPB Software (http://spb.com/press/pressreleases/2011/full-data-traffic-control-for-symbian-with-spb-wireless-monitor.html). Spb.com (2011-06-21). Retrieved on 2012-07-03.
[84] "iOS5 Features PC Free" (http://www.apple.com/ios/features.html#pcfree). Apple Inc.. 2011-06-11. . Retrieved 2011-09-07.
[85] Windows Phone 8 Devices Will Get OTA Updates and 18 Months of Firmware Support (http://gizmodo.com/5919970/windows-phone-8-devices-will-get-ota-updates-and-18-months-of-firmware-support)
[86] "BlackBerry Wireless Update" (http://www.youtube.com/watch?v=YgYYONd4eE8). youtube.com. 2010-04-05. . Retrieved 2011-09-07.
[87] "Updating the Firmware on your Symbian Phone: Two Methods Explained and Compared" (http://www.brighthub.com/mobile/symbian-platform/articles/33440.aspx). Brighthub.com. 2011-07-04. . Retrieved 2011-09-07.
[88] "N900 – Wireless Updating in Maemo 5" (http://www.youtube.com/watch?v=At9tJHmgnrE). youtube.com. 2009-10-20. . Retrieved 2011-09-07.
[89] "Does the iPad support multiple users?" (http://www.theipadguide.com/faq/does-ipad-support-multiple-users). The iPad Guide. 2010-04-11. . Retrieved 2011-09-07.
[90] "Android Issue 15030: Multi user support" (http://code.google.com/p/android/issues/detail?id=15030). Google.com. 2011-02-25. . Retrieved 2011-09-07.
[91] Multiple User Accounts Are Coming To Android: Miles Of Code Is Already In AOSP, And Some Of It Is Quietly Working On Devices Right Now (http://www.androidpolice.com/2012/07/30/multiple-user-accounts-are-coming-miles-of-code-is-already-in-aosp-and-some-of-it-is-quietly-working-on-devices-right-now/)
[92] "PreCentral Forums > webOS tablets > HP TouchPad > Multiple Users on Touchpad?" (http://forums.precentral.net/hp-touchpad/275059-multiple-users-touchpad.html). precentral.net. . Retrieved 2011-09-07.
[93] "MeeGo Feature 2682 – Add multiuser support" (https://bugs.meego.com/show_bug.cgi?id=2682). MeeGo.com. 2011-07-08. . Retrieved 2011-09-07.
[94] "iOS6 Guided Access" (http://www.apple.com/ios/ios6/#accessibility). Apple Inc.. 2012-06. . Retrieved 2012-06-26.
[95] Android Issue 7472: Guest Mode (http://code.google.com/p/android/issues/detail?id=7472). Code.google.com (2010-03-31). Retrieved on 2012-07-03.
[96] "MIUI Official Site" (http://en.miui.com/). En.miui.com. . Retrieved 2012-03-02.
[97] "iOS 4 New Features: New Location Services settings" (http://www.youtube.com/watch?v=XH0kvnFEi_0). youtube.com. 2010-06-20. . Retrieved 2011-09-07.
[98] "Notification Center" (http://www.apple.com/ios/features.html#notification). Apple Inc.. 2011-06-11. . Retrieved 2011-09-07.
[99] "Inside Apple's iOS 5: Notification Center, banner alerts" (http://www.appleinsider.com/articles/11/06/07/inside_ios_5_notification_center_banner_alerts.html). AppleInsider.com. 2011-06-07. . Retrieved 2011-09-07.
[100] Apple Adds New Privacy Settings in iOS 6 (http://www.iphonehacks.com/2012/06/ios-6-new-privacy-settings.html). Iphonehacks.com (2012-06-12). Retrieved on 2012-07-03.
[101] "Android Issue 3778: Feature request: Application permissions should be individually grantable by the user" (http://code.google.com/p/android/issues/detail?id=3778). Google.com. 2009-09-04. . Retrieved 2011-09-07.
[102] "New CM7 Nightly Allows Users To Deny Permissions For Applications" (http://www.droid-life.com/2011/05/23/new-cm7-nightly-allows-users-to-deny-permissions-for-applications/). Droid Life. 2011-05-23. . Retrieved 2011-09-07.
[103] "Mobile Application Security : WebOS Security – Permissions and User Controls" (http://programming4.us/mobile/3260.aspx). programming4.us. 2011-02-26. . Retrieved 2011-09-07.
[104] "Zune Marketplace requires user-confirmed permissions, similar to Android" (http://windowsphone7central.com/news/zune_marketplace_requires_userconfirmed_permissions_similar_to_android-10-11-10.php). windowsphone7central.com. 2010-11-10. . Retrieved 2011-09-07.
[105] Holder, Joseph (2010-12-31). "BlackBerry 101 – Application permissions" (http://crackberry.com/blackberry-101-application-permissions). crackberry.com. . Retrieved 2011-09-07.

[106] "How To Set Up iOS Parental Controls" (http://www.internetsafetyproject.org/wiki/parental-controls-iphone-ipod-touch-and-ipad). Internet Safety Project. 2011-06-03. . Retrieved 2011-09-07.

[107] "Android Parental Control Apps" (http://www.tomsguide.com/us/Parental-Controls-Android-iOS-Twitter-Facebook,review-1680-6.html). tomsguide.com. 2011-07-28. . Retrieved 2011-09-07.

[108] "Find My iPhone" (http://www.apple.com/iphone/features/find-my-iphone.html). Apple Inc.. 2009-06-10. . Retrieved 2011-09-07.

[109] "Prey Project" (http://preyproject.com/). Fork Ltd. 2011-10-18. . Retrieved 2011-10-18. and [Google_Latitude|Google Latitude]

[110] Mathews, Lee (2010-09-04). "Windows Phone 7 remote lock and wipe, find my phone features pictured" (http://downloadsquad.switched.com/2010/09/04/windows-phone-7-remote-lock-and-wipe-features-pictured/). Downloadsquad.switched.com. . Retrieved 2012-03-02.

[111] "Microsoft Exchange Information Services and Security Policies Supported by Android 2.2 and 2.3" (http://static.googleusercontent.com/external_content/untrusted_dlcp/www.google.com/en/us/help/hc/pdfs/mobile/ExchangeAndAndroid2.2and2.3-003.pdf). Google. .

[112] "Palm webOS gains enterprise features" (http://www.networkworld.com/news/2009/072409-palm-webos-gains-enterprise-features.html?page=1). Network World. 2009-07-24. . Retrieved 2011-09-13.

[113] "iPhone Technical Specifications" (http://www.apple.com/iphone/specs.html). Apple Inc.. . Retrieved 2011-09-07.

[114] "iOS: Using international keyboards" (http://support.apple.com/kb/HT4509). Apple Inc.. 2011-02-02. . Retrieved 2011-09-07.

[115] "iPhone and iPod touch: Supported languages for Voice Control and VoiceOver" (http://support.apple.com/kb/HT3562). Apple Inc.. 2010-09-09. . Retrieved 2011-09-07.

[116] Android Issue 3732: Search Improve contact search (http://code.google.com/p/android/issues/detail?id=3732) Android Issue 3393: No Unicode support on SMS sending (http://code.google.com/p/android/issues/detail?id=3393) Android Issue 9199: SQLite3 Unicode Case not working (only ASCII case works) (include ICU support) (http://code.google.com/p/android/issues/detail?id=9199)

[117] "iOS 4 features: Spell-check and text replace" (http://www.tipb.com/2010/07/13/ios-4-features-spellcheck-text-replace/). TiPb.com. 2010-07-13. . Retrieved 2011-09-07.

[118] {{cite web|url=http://code.google.com/p/android/issues/detail?id=4939 |title=Support spell-checking when typing with the hardware keyboard

[119] "Android Issue 10332: Retrospective spell checking on all input fields as an OS feature (input device agnostic)" (http://code.google.com/p/android/issues/detail?id=10332). Google.com. 2010-08-10. . Retrieved 2011-09-07.

[120] "Windows Phone 7 review" (http://www.engadget.com/2010/10/20/windows-phone-7-review/). Engadget. 2010-10-20. . Retrieved 2011-09-07.

[121] "New in iOS 5: Dictionary" (http://www.idownloadblog.com/2011/06/07/ios-5-dictionary/). idownloadblog.com. 2011-06-07. . Retrieved 2011-09-07.

[122] "Edit Android built (not user) dictionary" (http://androidforums.com/android-lounge/413766-edit-android-built-not-user-dictionary.html). .

[123] "iOS 5: OMW! How To Create Keyboard Text Shortcuts" (http://iostips.net/ios-5-omw-how-create-keyboard-text-shortcuts/). iostips.net. . Retrieved 2011-09-07.

[124] "Working with Text Assist" (http://kb.hpwebos.com/wps/portal/kb2/common/article/70980_en.html). HP. . Retrieved 2011-09-13.

[125] Kyle Barrow (2009-09-30). "iPhone Emoji" (http://pukupi.com/post/1964). pukupi.com. . Retrieved 2011-09-07.

[126] Android Issue 8375: Add emoji support to android (http://code.google.com/p/android/issues/detail?id=8375). Code.google.com (2010-05-17). Retrieved on 2012-07-03.

[127] Go Sms Pro (https://play.google.com/store/apps/details?id=com.jb.gosms). Play.google.com (2012-06-21). Retrieved on 2012-07-03.

[128] "Windows Phone around the world: Language support in Mango" (http://windowsteamblog.com/windows_phone/b/windowsphone/archive/2011/07/06/windows-phone-around-the-world-language-support-in-mango.aspx). Windowsteamblog.com. 2011-05-25. . Retrieved 2012-03-02.

[129] "Accessibility – iPhone" (http://www.apple.com/accessibility/iphone/vision.html). Apple Inc.. . Retrieved 2011-09-07.

[130] "iPhone 4 – Accessibility features are built right in" (http://www.apple.com/iphone/features/accessibility.html). Apple Inc.. 2010-06-21. . Retrieved 2011-09-07.

[131] Pogue, David. (2011-11-10) Apple's AssistiveTouch Helps the Disabled Use a Smartphone – NYTimes.com (http://pogue.blogs.nytimes.com/2011/11/10/apples-assistivetouch-helps-the-disabled-use-a-smartphone/?nl=technology&emc=cta2&pagewanted=all). Pogue.blogs.nytimes.com. Retrieved on 2012-07-03.

[132] Bonnie Cha (27 October 2010). "Top 5 things I dislike about Windows Phone 7" (http://www.cnet.com/8301-17918_1-20020922-85.html). CNET. .

[133] "iPhone 3.0 spotlight – bookmarks?" (http://forums.macrumors.com/showthread.php?t=711973). macrumors.com. . Retrieved 2011-09-07.

[134] Issue 2518 (http://code.google.com/p/android/issues/detail?id=2518). code.google.com (2009-04-27).

[135] How do I search for a SMS/Text message? – Microsoft Answers (http://answers.microsoft.com/en-us/winphone/forum/wp7-wptips/how-do-i-search-for-a-smstext-message/f448b4c3-315b-e011-8dfc-68b599b31bf5). Answers.microsoft.com (2011-09-06). Retrieved on 2012-07-03.

[136] Search Improved With Mango, But Still Lacking In Features – myWindowsPhone (http://www.mywindowsphone.net/102/search-improved-with-mango-but-still-lacking-in-features). Mywindowsphone.net (2011-09-30). Retrieved on 2012-07-03.

[137] Can't Search for music in Music & video Hub by using keywords? – Microsoft Answers (http://answers.microsoft.com/en-us/winphone/forum/wp7-wpmusic/cant-search-for-music-in-music-video-hub-by-using/b75fab53-2874-e011-8dfc-68b599b31bf5). Answers.microsoft.com (2011-09-06). Retrieved on 2012-07-03.
[138] iOS 6: minor tweaks that you might like (http://www.phonearena.com/news/iOS-6-minor-tweaks-that-you-might-like_id31135). Phonearena.com. Retrieved on 2012-07-03.
[139] "Android Issue 3732: Only search Contacts name field but not any other field" (http://code.google.com/p/android/issues/detail?id=3732). Google.com. 2009-08-31. . Retrieved 2011-09-07.
[140] "WP7 on HTC Mozart – Unable to search for additional Contact fields – by design or bug?" (http://social.answers.microsoft.com/Forums/en-US/windowsphone7/thread/b29e7b1a-c870-4061-bad7-9669187a465f). Microsoft. 2010-11-11. . Retrieved 2011-09-07.
[141] "Notes search not available in BB Torch" (http://supportforums.blackberry.com/t5/BlackBerry-Torch-9800-smartphone/Notes-search-not-available-in-BB-Torch/m-p/675645). Research In Motion. 2010-12-11. . Retrieved 2011-09-07.
[142] Torch Universal Search – BlackBerry Forums at CrackBerry.com (http://forums.crackberry.com/blackberry-torch-9800-f209/torch-universal-search-585621/). Forums.crackberry.com (2011-02-13). Retrieved on 2012-07-03.
[143] "Android Issue 1273: IP Proxy Settings for Wifi Network" (http://code.google.com/p/android/issues/detail?id=1273). Google.com. 2008-11-12. . Retrieved 2011-09-07.
[144] "Android Issue 4386: Add proxy options for each individual internet connection" (http://code.google.com/p/android/issues/detail?id=4386). Google.com. 2009-10-27. . Retrieved 2011-09-07.
[145] "webOS Software : No proxy support when using WiFi on Pre" (http://forums.palm.com/t5/webOS-Software/No-proxy-support-when-using-WiFi-on-Pre/td-p/208059). palm.com. . Retrieved 2011-09-07.
[146] "iOS: Understanding data protection" (http://support.apple.com/kb/HT4175). Apple Inc.. 2011-10-14. . Retrieved 2011-10-20.
[147] Recover passwords protecting iPhone/iPod and BlackBerry backups (http://www.elcomsoft.com/eppb.html). Elcomsoft.com. Retrieved on 2012-07-03.
[148] Android Issue 3748: Add support for partition/block device encryption (http://code.google.com/p/android/issues/detail?id=3748). Code.google.com (2009-09-01). Retrieved on 2012-07-03.
[149] "Microsoft Windows Mobile encryption" (http://msdn.microsoft.com/en-us/library/bb416357.aspx). Microsoft. 2010-04-08. . Retrieved 2011-09-07.
[150] MS teases Windows Phone 8 enterprise features: Company Hub, encryption, secure boot, IT management -- Engadget (http://www.engadget.com/2012/06/20/windows-phone-8-enterprise-features/)
[151] "BlackBerry Stored Data Security" (http://us.blackberry.com/ataglance/security/features.jsp#tab_tab_stored_data). Research In Motion. . Retrieved 2011-09-07.
[152] Android Issue 11211: Android too insecure – Encryption of the SDcard is crucial (http://code.google.com/p/android/issues/detail?id=11211). Code.google.com (2010-09-14). Retrieved on 2012-07-03.
[153] How to secure your BlackBerry « BlackBerryInsight – BlackBerry News & Information (http://www.blackberryinsight.com/2007/12/04/how-to-secure-your-blackberry/). Blackberryinsight.com. Retrieved on 2012-07-03.
[154] "Does iCloud encrypt files before uploading?" (https://discussions.apple.com/thread/3103306). Apple Inc.. 2011-06-07. . Retrieved 2011-09-07.
[155] "Google Admits Handing over European User Data to US Intelligence Agencies" (http://news.softpedia.com/news/Google-Admits-Handing-over-European-User-Data-to-US-Intelligence-Agencies-215740.shtml). softpedia.com. 2011-08-08. . Retrieved 2011-09-07.
[156] Microsoft: European cloud data may not be immune to the Patriot Act (http://www.engadget.com/2011/06/30/microsoft-european-cloud-data-may-not-be-immune-to-the-patriot/). Engadget.com (2011-06-30). Retrieved on 2012-07-03.
[157] Are There Any Smartphones That Respect Privacy? – Slashdot (http://yro.slashdot.org/story/11/11/22/0015223/are-there-any-smartphones-that-respect-privacy). Yro.slashdot.org. Retrieved on 2012-07-03.
[158] Windows Phone 7 is a good solution – Are There Any Smartphones That Respect Privacy? – Slashdot (http://yro.slashdot.org/comments.pl?sid=2537718&cid=38131168). Yro.slashdot.org. Retrieved on 2012-07-03.
[159] How do I disable skydrive – Microsoft Answers (http://answers.microsoft.com/en-us/winphone/forum/wp7-wptips/how-do-i-disable-skydrive/c31aed58-bd64-43a3-a2cd-2fb54752f4d5?msgId=16276103-b77c-4e2d-902b-bd80936bf814). Answers.microsoft.com (2011-09-06). Retrieved on 2012-07-03.
[160] Want Privacy? Get your own BES – Are There Any Smartphones That Respect Privacy? – Slashdot (http://yro.slashdot.org/comments.pl?sid=2537718&cid=38131262). Yro.slashdot.org (2011-11-21). Retrieved on 2012-07-03.
[161] "iPhone in Business – Integration" (http://www.apple.com/iphone/business/integration/). Apple Inc.. . Retrieved 2011-09-07.
[162] "Catching AuthTokens in the Wild – The Insecurity of Google's ClientLogin Protocol" (http://www.uni-ulm.de/in/mi/mitarbeiter/koenings/catching-authtokens.html). uni-ulm.de. 2011-06-15. . Retrieved 2011-09-07.
[163] "Productivity takes a big step forward in Mango" (http://windowsteamblog.com/windows_phone/b/windowsphone/archive/2011/05/16/productivity-takes-a-big-step-forward-in-mango.aspx). windowsteamblog.com. 2011-05-16. . Retrieved 2011-09-07.
[164] "Standard BlackBerry encryption" (http://docs.blackberry.com/en/admin/deliverables/12873/Standard_BlackBerry_message_encryption_193608_11.jsp). Research In Motion. . Retrieved 2011-09-07.
[165] "Android Issue 66: OpenSync desktop synchronisation support" (http://code.google.com/p/secrets-for-android/issues/detail?id=66). Google.com. 2010-05-04. . Retrieved 2011-09-07.

[166] "Application – HTC Sync 3.0 for HTC Desire HD, Desire Z, Legend, Desire, Wildfire, Aria, Gratia & Android 2.1 Hero" (http://www.htc.com/europe/SupportViewNews.aspx?dl_id=982&news_id=769). HTC.com. 2010-11-08. . Retrieved 2011-09-07.

[167] "Palm needs to develop a Palm Pre desktop sync application" (http://forums.palm.com/t5/Synergy-webOS/Palm-needs-to-develop-a-Palm-Pre-desktop-sync-application/td-p/209510). palm.com. 2009-06-09. . Retrieved 2011-09-07.

[168] Lein, Adam Z (2011-02-25). "How to Sync Windows Phone 7 with Outlook" (http://pocketnow.com/windows-phone/how-to-sync-windows-phone-7-with-outlook). pocketnow.com. . Retrieved 2011-09-07.

[169] "Petition Microsoft to reinstate usb sync for Outlook" (http://answers.microsoft.com/en-us/winphone/forum/wp7-sync/petition-microsoft-to-reinstate-usb-sync-for/2d288794-2d56-e011-8dfc-68b599b31bf5). Microsoft. 2011-03-24. . Retrieved 2011-09-07.

[170] "Maemo Sync" (http://wiki.maemo.org/Sync). maemo.org. 2011-08-24. . Retrieved 2011-09-07.

[171] "iTunes: Backing up, updating, and restoring your iPhone, iPad, or iPod touch software" (http://support.apple.com/kb/ht1414). Apple Inc.. 2010-08-13. . Retrieved 2011-09-07.

[172] "Android Issue 1106: No way to backup or restore on-device data (sms, most application data, installed apps and settings)" (http://code.google.com/p/android/issues/detail?id=1106). Google.com. 2008-10-28. . Retrieved 2011-09-07.

[173] "Titanium Backup for Android – Official Site" (http://matrixrewriter.com/android/). matrixrewriter.com. . Retrieved 2011-09-07.

[174] "webOS backup utility in the works" (http://www.precentral.net/webos-backup-utility-works). precentral.net. 2009-11-18. . Retrieved 2011-09-07.

[175] "Complete Windows Phone 7 Backup" (http://answers.microsoft.com/en-us/winphone/forum/wp7-sync/complete-windows-phone-7-backup/4d38a00c-917f-497c-96b9-b60e29feb5f0). Microsoft. 2010-12-06. . Retrieved 2011-09-07.

[176] "How to back up the data on a BlackBerry smartphone" (http://www.blackberry.com/btsc/search.do?cmd=displayKC&docType=kc&externalId=KB12487). Research In Motion. 2011-01-06. . Retrieved 2011-09-07.

[177] BlackBerry OS 7 Device Switch (http://www.youtube.com/watch?v=ILQamIoJ4pg). Youtube.com (2011-06-16). Retrieved on 2012-07-03.

[178] "Is it possbile to backup phone including apps" (http://www.badaforums.net/forums/samsung-wave-forum/possbile-backup-phone-including-apps-t370.html). badaforums.net. 2010-08-08. . Retrieved 2011-09-07.

[179] "Android Issue 2907: Google Bookmarks should be integrated with the browser" (http://code.google.com/p/android/issues/detail?id=2907). Google.com. 2009-06-07. . Retrieved 2011-09-07.

[180] can i sync favourites from IE PC and IE windows phone 7? – Microsoft Answers (http://answers.microsoft.com/en-us/winphone/forum/wp7-sync/can-i-sync-favourites-from-ie-pc-and-ie-windows/b39a2031-e888-48ef-9907-e6f3d02d171b). Answers.microsoft.com (2011-09-06). Retrieved on 2012-07-03.

[181] windows phone & sync text messages – Microsoft Answers (http://answers.microsoft.com/en-us/winphone/forum/wp7-sync/windows-phone-sync-text-messages/2511cb77-9073-4a86-8df7-a950c11fd145). Answers.microsoft.com (2011-09-06). Retrieved on 2012-07-03.

[182] MyPhone: Backup Email settings – Microsoft Answers (http://answers.microsoft.com/en-us/winphone/forum/wp6n-wpmyphone/myphone-backup-email-settings/3cf4a396-66f7-4c72-a536-d4b2b982ddc4). Answers.microsoft.com (2011-09-06). Retrieved on 2012-07-03.

[183] Nilay Patel (2009-03-17). "iPhone finally gets copy and paste!" (http://www.engadget.com/2009/03/17/iphone-finally-gets-copy-and-paste/). *Engadget*. .

[184] "Android Issue 3190: Improve copy-paste in Browser/WebView" (http://code.google.com/p/android/issues/detail?id=3190). Google.com. 2009-07-06. . Retrieved 2011-09-07.

[185] "Microsoft's Windows Phone 7 'NoDo' update starts rolling out, brings copy / paste and other fun additions" (http://www.engadget.com/2011/03/22/microsofts-windows-phone-7-nodo-update-starts-rolling-out-br/). Engadget. 2011-03-22. . Retrieved 2011-09-07.

[186] "Copy and paste coming to Windows Phone 7 in 'early 2011' (update)" (http://www.engadget.com/2010/10/11/copy-and-paste-coming-to-windows-phone-7-in-early-2011/). Engadget. 2010-10-11. . Retrieved 2011-09-07.

[187] "Does Bada OS have Cut, Copy & Paste?" (http://forum2.mobile-review.com/showthread.php?t=92645). mobile-review.com. 2011-07-01. . Retrieved 2011-09-07.

[188] Phat^Trance (2009-05-23). "iPhone: Undo Typing – Cool feature in iPhone 3.0" (http://dailymobile.se/2009/05/23/iphone-undo-typing-nice-feature-in-iphone-30/). dailymobile. .

[189] "Android Issue 6458: Enhancement: Undo" (http://code.google.com/p/android/issues/detail?id=6458). Google.com. 2010-02-04. . Retrieved 2011-09-07.

[190] "Android 4.0 (Ice Cream Sandwich) SDK and ADT 14 Released" (http://www.androidpolice.com/2011/10/18/android-4-0-ice-cream-sandwich-sdk-and-adt-14-released/). Android Police. October 19, 2011. . Retrieved October 19, 2011.

[191] Miniman, Brandon. "Visual Voicemail in Windows Phone Mango (Video)" (http://pocketnow.com/windows-phone/visual-voicemail-in-windows-phone-mango-video). Pocketnow.com. . Retrieved 2012-03-02.

[192] "Smart Visual Voicemail for Nokia Series 40 devices" (http://www.communology.com/index.php?id=visual_voicemail). communology.com. . Retrieved 2012-08-02.

[193] iOS 4 on the iPhone has Improved Call History (http://www.simplemobilereview.com/ios-4-on-the-iphone-has-improved-call-history/). Simplemobilereview.com (2010-09-02). Retrieved on 2012-07-03.

[194] A Week with the Android G1 Smartphone – Day 3 & 4 — The Gadgeteer (http://the-gadgeteer.com/2009/01/29/a-week-with-the-android-g1-smartphone-day-3-4/). The-gadgeteer.com. Retrieved on 2012-07-03.

[195] Call History – Call Length – Microsoft Answers (http://answers.microsoft.com/en-us/winphone/forum/wp7-wptips/call-history-call-length/a83d9fd3-cc69-4392-801a-605939cbb7a3). Answers.microsoft.com (2011-09-06). Retrieved on 2012-07-03.

[196] About call logs – User Guide – BlackBerry Bold 9780 Smartphone – 6.0 (http://docs.blackberry.com/en/smartphone_users/deliverables/22178/About_call_logs_26291_11.jsp). Docs.blackberry.com. Retrieved on 2012-07-03.

[197] How can I set a second mobile number in a Windows Live contact? – Microsoft Answers (http://answers.microsoft.com/en-us/winphone/forum/wp7-wppeople/how-can-i-set-a-second-mobile-number-in-a-windows/0b8e1c44-1087-40cd-8cf0-18367682f3de). Answers.microsoft.com (2011-09-06). Retrieved on 2012-07-03.

[198] App Store – Groups: SMS, Mail and Manage Contacts (http://itunes.apple.com/es/app/groups-sms-mail-manage-contacts/id300891442). Itunes.apple.com (2011-12-19). Retrieved on 2012-07-03.

[199] Android 4 Breaks New Ground (Part 4) – infoSync (http://www.infosyncworld.com/reviews/cell-phones/android-4-breaks-new-ground-(part-4)/12197.html)

[200] Manage your contacts with groups on your android – Clix Group Manager (http://www.youtube.com/watch?v=8ajmcejVf88). Youtube.com (2009-09-30). Retrieved on 2012-07-03.

[201] Groups on Windows Phone 7.5 (http://www.youtube.com/watch?v=hqw_GLXgs4s). Youtube.com (2011-09-01). Retrieved on 2012-07-03.

[202] "Android Issue 4575: Phone number linking to dialer" (http://code.google.com/p/android/issues/detail?id=4575#56). Google.com. 2009-11-07. . Retrieved 2011-09-07.

[203] "Google Wallet - Android Appstore" (https://play.google.com/store/apps/details?id=com.google.android.apps.walletnfcrel&hl=en). Google.com. 2012-06-01. . Retrieved 2012-06-01.

[204] http://www.phonearena.com/news/Windows-Phone-8-the-new-features_id31466#11-Wallet-Hub

[205] Sande, Steven (2009-06-22). "Inside iPhone 3.0: The address data detector" (http://www.tuaw.com/2009/06/22/inside-iphone-3-0-the-address-data-detector/). tuaw.com. . Retrieved 2011-09-07.

[206] "Android Issue 19527: Calendar App: Ability to click on location in Event view" (http://code.google.com/p/android/issues/detail?id=19527). Google.com. 2011-08-24. . Retrieved 2011-09-07.

[207] "Street addresses not showing up as links from gmail mobile" (http://www.google.ru/support/forum/p/Google+Mobile/thread?tid=237056226cd02b10&hl=en). Google.ru. 2011-06-26. . Retrieved 2011-09-07.

[208] Fried, Ina (2010-03-18). "A look at Windows Phone's 'smart links' | Beyond Binary – CNET News" (http://news.cnet.com/8301-13860_3-20000737-56.html). News.cnet.com. . Retrieved 2012-03-02.

[209] "How to Create Calendar Events from Emails in iPhone Mail" (http://email.about.com/od/iphonemailtips/qt/How-to-Create-Calendar-Events-from-Emails-in-iPhone-Mail.htm). email.about.com. 2010-06-23. . Retrieved 2011-09-07.

[210] Importing ICS Files into the Calendar – Instant Expert: Secrets & Features of iOS 4.2 (http://www.ilounge.com/index.php/articles/comments/instant-expert-secrets-features-of-ios-4.2/). iLounge Article. Retrieved on 2012-07-03.

[211] Android Issue 1257: Calendar should support opening iCal .ics files (http://code.google.com/p/android/issues/detail?id=1257). Code.google.com (2008-11-11). Retrieved on 2012-07-03.

[212] iCalParse/CalendarSync (version 7.6) importing iCalendar informations on Android – xda-developers (http://forum.xda-developers.com/showthread.php?t=905213). Forum.xda-developers.com. Retrieved on 2012-07-03.

[213] Windows Mobile importing iCalendar(iCalParse 2.1.1) – xda-developers (http://forum.xda-developers.com/showthread.php?t=544181). Forum.xda-developers.com. Retrieved on 2012-07-03.

[214] how do i open an ical appointment in windows phone 7 – Microsoft Answers (http://answers.microsoft.com/en-us/winphone/forum/wp7-wppeople/how-do-i-open-an-ical-appointment-in-windows-phone/b323fe76-27e0-490d-9ac3-37a6fd0b1eb6). Answers.microsoft.com (2011-09-06). Retrieved on 2012-07-03.

[215] Add an appointment to your calendar from a text message – BlackBerry Curve 9350/9360/9370 Smartphones – User Guide – BlackBerry Curve Series – 7.0 (http://docs.blackberry.com/en/smartphone_users/deliverables/33213/Add_an_appt_to_your_calendar_from_a_text_msg_60_1052979_11.jsp). Docs.blackberry.com. Retrieved on 2012-07-03.

[216] SyncMate - Establish Connection with Nokia S40 Devices (http://wiki.eltima.com/user-guides/syncmate-macos/nokia.html). wiki.eltima.com. Retrieved on 2012-08-02.

[217] "Google Chrome for Android" (https://www.google.com/intl/en/chrome/browser/mobile/android.html). Google.com. .

[218] "Android Issue 8880: Keep open URLs/windows on shutdown or crash" (http://code.google.com/p/android/issues/detail?id=8880). Google.com. 2010-06-06. . Retrieved 2011-09-07.

[219] Android all_search_engines.xml (https://www.codeaurora.org/git/projects/qrd-gb-dsds-7225/repository/revisions/f6156a77044113680d038934b06bde333a1e69ea/entry/packages/apps/Browser/res/values/all_search_engines.xml). Codeaurora.org. Retrieved on 2012-07-03.

[220] "Tip: Just Type: Add website searches" (http://www.precentral.net/just-type-add-website-searches). PreCentral. 2010-06-23. . Retrieved 2011-05-16.

[221] By: snakechia. "Tip: T-Mobile HD7 Mango gets Search Engine option | wpcentral | Windows Phone News, Forums, and Reviews" (http://www.wpcentral.com/t-mobile-hd7-mango-gets-search-engine-option). wpcentral. . Retrieved 2012-03-02.

[222] "iOS 4.2 features: Find text on Safari web page" (http://www.tipb.com/2010/09/16/ios-42-features-find-text-safari-web-page/). TiPb.com. 2010-09-16. . Retrieved 2011-09-07.

[223] "Cupcake Adds Camcorder, Touchscreen Keyboard, and More to Android" (http://smarterware.org/1744/cupcake-adds-video-camera-touchscreen-keyboard-and-more-to-android). smarterware.org. 2009-05-27. . Retrieved 2011-09-07.

[224] Is Find on Page feature being removed from IE9 Mango?! (http://answers.microsoft.com/en-us/winphone/forum/wp7-wpfeatures/is-find-on-page-feature-being-removed-from-ie9/22e6b0fc-449c-4a85-9ab0-9d4c134a4a85?msgId=7f24a077-c916-47c7-89f7-be8cea73f6a1) answers.microsoft.com (2011-06-22).

[225] "BlackBerry Torch 9800 Smartphone User Guide Version: 6.0" (http://docs.blackberry.com/en/smartphone_users/deliverables/18579/BlackBerry_Torch_9800_Smartphone-User_Guide-T643442-941426-0810050917-001-6.0-US.pdf) (PDF). Research In Motion. . Retrieved 2011-09-07.

[226] "iPhone Tip: Save Images in iPhone's Safari browser and Mail App" (http://www.iphonehacks.com/2008/07/save-images.html). iphonehacks.com. . Retrieved 2011-09-07.

[227] How do you save a picture from the web page or googled images on – Microsoft Answers (http://answers.microsoft.com/en-us/winphone/forum/wp7-wppics/how-do-you-save-a-picture-from-the-web-page-or/222970d0-b96c-e011-8dfc-68b599b31bf5?msgId=c302e2e8-a56d-e011-8dfc-68b599b31bf5). Answers.microsoft.com (2011-09-06). Retrieved on 2012-07-03.

[228] Apple – iPhone – Tips and Tricks (http://www.apple.com/iphone/tips/). Apple.com. Retrieved on 2012-07-03.

[229] "Adobe brings PDF support to Windows Phone 7" (http://wmpoweruser.com/adobe-brings-pdf-support-to-windows-phone-7/). WMPoweruser. . Retrieved 2012-03-02.

[230] iOS 6: All the New Features Coming Soon to Your iPhone and iPad (http://gizmodo.com/5917359/ios-6-new-features-coming-soon-to-your-iphone-and-ipad). Gizmodo.com (2012-06-11). Retrieved on 2012-07-03.

[231] App Store – Instapaper (http://itunes.apple.com/us/app/instapaper/id288545208). Itunes.apple.com (2012-06-02). Retrieved on 2012-07-03.

[232] "First Look at the Highly Polished Android 4.0, "Ice Cream Sandwich"" (http://lifehacker.com/5851162/first-look-at-the-highly-polished-android-40-ice-cream-sandwich). Lifehacker. October 19, 2011. . Retrieved October 19, 2011.

[233] "Save a web page to the main messages application – User Guide – BlackBerry Bold 9700 Smartphone – 5.0" (http://docs.blackberry.com/en/smartphone_users/deliverables/11298/Save_a_web_page_to_a_message_list_30041_11.jsp). Research In Motion. . Retrieved 2011-09-07.

[234] Download any file on your iPhone with iDownloader Pro (http://www.iphoneactive.com/2011/12/download-any-file-on-your-iphone-with.html). Iphone Active (2011-12-02). Retrieved on 2012-07-03.

[235] Android Issue 2450: Download any type of file (http://code.google.com/p/android/issues/detail?id=2450). Code.google.com (2009-04-17). Retrieved on 2012-07-03.

[236] Android Browser File Types – Android Wiki (http://android-dls.com/wiki/index.php?title=Android_Browser_File_Types). Android-dls.com (2011-12-11). Retrieved on 2012-07-03.

[237] Add-In: WHS Phone for Windows Phone 7 Version 1.3 (http://www.mswhs.com/2011/05/add-in-whs-phone-for-windows-phone-7-version-1-3/). Mswhs.com (2011-05-25). Retrieved on 2012-07-03.

[238] "iOS 5 Safari" (http://www.apple.com/ios/features.html#safari). Apple Inc.. 2011-06-11. . Retrieved 2011-09-07.

[239] Inside Apple's iOS 5: Safari Reader, private browsing, tabs on iPad (http://www.appleinsider.com/articles/11/06/07/inside_apples_ios_5_safari_reader_private_browsing_tabs_on_ipad/). Appleinsider.com (2011-06-07). Retrieved on 2012-07-03.

[240] iOS browsers that allow font scaling and text reflow (http://www.456bereastreet.com/archive/201101/ios_browsers_that_allow_font_scaling_and_text_reflow/). 456 Berea Street (2011-01-05). Retrieved on 2012-07-03.

[241] Browser Text Reflow – WP7 Internet Explorer vs Android (http://www.youtube.com/watch?v=LpQhXZH1_1Y). Youtube.com (2011-12-07). Retrieved on 2012-07-03.

[242] "Palm, please add Text Reflow to the Web Browser" (http://forums.precentral.net/palm-pre/253307-palm-please-add-text-reflow-web-browser.html). precentral.net. 2010-06-30. . Retrieved 2011-09-07.

[243] Cha, Bonnie (2010-08-05). "BlackBerry OS 6 | Download Product Review" (http://download.cnet.com/8301-20_4-10475434-10084490.html). Download.com. . Retrieved 2011-09-07.

[244] How do you like the new reader mode for the browser – BlackBerry Forums at CrackBerry.com (http://forums.crackberry.com/blackberry-playbook-f222/how-do-you-like-new-reader-mode-browser-661969/). Forums.crackberry.com (2011-10-19). Retrieved on 2012-07-03.

[245] "'Safari Upload Enabler' Lets You Upload Files from Your iPhone" (http://www.idownloadblog.com/2011/08/03/safari-upload-enabler/). www.idownloadblog.com. 2011-08-03. . Retrieved 2011-11-04.

[246] "Android Issue 2519: Browser should support file uploads (input type="file")" (http://code.google.com/p/android/issues/detail?id=2519). Google.com. 2009-04-27. . Retrieved 2011-09-07.

[247] Android Issue 15571: DalvikVM Crashes when upload 50MB+ file to youtube through Browser (http://code.google.com/p/android/issues/detail?id=15571). Code.google.com (2011-03-16). Retrieved on 2012-07-03.

[248] "Safari Web Content Guide: Laying Out Forms" (http://developer.apple.com/library/iOS/#documentation/AppleApplications/Reference/SafariWebContent/DesigningForms/DesigningForms.html#//apple_ref/doc/uid/TP40006512-SW5). Apple Inc.. 2010-12-18. . Retrieved 2011-09-07.

[249] "How can i switch to next field in web form using Android browser?" (http://android.stackexchange.com/questions/11018/how-can-i-switch-to-next-field-in-web-form-using-android-browser). android.stackexchange.com. 2011-07-09. . Retrieved 2011-09-07.

[250] "Windows Phone 7 – Form Navigation" (http://stackoverflow.com/questions/6739528/form-navigation). stackoverflow.com. 2011-07-19. . Retrieved 2011-09-07.

[251] Joe White (2011-06-08). "Safari In iOS 5 Adds "Private Browsing" – Easily Protect Your Privacy While Surfing The Web" (http://appadvice.com/appnn/2011/06/safari-ios-5-adds-private-browsing-protect-privacy). AppAdvice. . Retrieved 2011-09-07.

[252] Android updates guide: All the features of every version (http://reviews.cnet.co.uk/mobile-phones/android-updates-guide-all-the-features-of-every-version-50003779/). Reviews.cnet.co.uk. Retrieved on 2012-07-03.

[253] "cyanogenmod" (http://www.cyanogenmod.com/). . Retrieved October 19, 2011.

[254] "Metro Browser 2.5 video review" (http://wmpoweruser.com/metro-browser-2-5-video-review/). WMPoweruser. . Retrieved 2012-03-02.

[255] How to create offline webapps on the iPhone | The CSS Ninja – All things CSS, JavaScript & HTML (http://www.thecssninja.com/javascript/how-to-create-offline-webapps-on-the-iphone). Thecssninja.com. Retrieved on 2012-07-03.

[256] Safari Web Content Guide: Configuring Web Applications (http://developer.apple.com/library/IOS/#documentation/AppleApplications/Reference/SafariWebContent/ConfiguringWebApplications/ConfiguringWebApplications.html). Developer.apple.com. Retrieved on 2012-07-03.

[257] Apple – Web apps (http://www.apple.com/webapps/whatarewebapps.html). Apple.com. Retrieved on 2012-07-03.

[258] iphone – Android Offline Webapp Resources – Stack Overflow (http://stackoverflow.com/questions/4321622/android-offline-webapp-resources). Stackoverflow.com. Retrieved on 2012-07-03.

[259] "iphone – How to change the language of the App store (and iTunes)? – Apple – Stack Exchange" (http://apple.stackexchange.com/questions/3897/how-to-change-the-language-of-the-app-store-and-itunes). apple.stackexchange.com. 2011-07-09. . Retrieved 2011-09-07.

[260] "Apple Releases Its Rules for App Store Censoring" (http://www.campusprogress.org/articles/apple_releases_its_rules_for_app_store_censoring/). campusprogress.org. 2010-09-10. . Retrieved 2011-09-07.

[261] "webOS Billing countries" (http://forum.xda-developers.com/archive/index.php/t-990047.html). xda-developers.com. 2011-03-12. . Retrieved 2011-09-07.

[262] "Impostah – PreCentral Forums" (http://forums.precentral.net/webos-internals/270316-impostah.html). precentral.net. 2010-12-05. . Retrieved 2011-09-07.

[263] windows phone 7 marketplace countries – Microsoft Answers (http://answers.microsoft.com/en-us/winphone/forum/wp6n-wpmarketplace/windows-phone-7-marketplace-countries/70d1f8e5-2966-4a26-a8cc-c0bf8477756d). Answers.microsoft.com (2011-09-06). Retrieved on 2012-07-03.

[264] Android Issue 8261: Unified Inbox (Gmail & other mailboxes in 1 place) (http://code.google.com/p/android/issues/detail?id=8261)

[265] K-9 Mail - Android-apps op Google Play (https://play.google.com/store/apps/details?id=com.fsck.k9&). Play.google.com. Retrieved on 2012-07-03.

[266] "Live from Microsoft's Windows Phone VIP preview event!" (http://www.engadget.com/2011/05/24/live-from-microsofts-windows-phone-vip-preview-event/). Engadget. 2011-05-24. . Retrieved 2011-09-07.

[267] "iOS 5 Notification Center" (http://www.apple.com/ios/features.html#notification). Apple Inc.. 2011-06-11. . Retrieved 2011-09-07.

[268] Wesley Fenlon (2011-03-01). "iOS vs Android vs WebOS: Who's Doing Mobile Notifications Right?" (http://www.tested.com/news/ios-vs-android-vs-webos-whos-doing-mobile-notifications-right/1918/). tested.com. . Retrieved 2011-09-07.

[269] Palm Pre Keynote CES 2009 (8/13) – webOS Notifications (http://www.youtube.com/watch?v=zplVSI0M93o). Youtube.com (2009-02-21). Retrieved on 2012-07-03.

[270] "The Windows Phone 7 Review" (http://www.anandtech.com/print/3982). AnandTech.com. 2010-10-20. . Retrieved 2011-09-07.

[271] Create, edit, or delete an appointment | Windows Phone 7 (http://www.microsoft.com/windowsphone/en-us/howto/wp7/people/create-edit-or-delete-an-appointment.aspx). microsoft.com

[272] "BlackBerry Is Ahead of the Game in Managing Notifications" (http://baconanthego.com/2011/05/01/blackberry-is-ahead-of-the-game-in-managing-notifications/). Bacon On The Go. 2011-05-01. . Retrieved 2011-09-07.

[273] Nguyen, Chuong (2011-07-08). "Symbian Belle Drop Down Notifications Reminiscent of Samsung's TouchWiz Notification Menu" (http://www.gottabemobile.com/2011/07/08/symbian-belle-drop-down-notifications-reminiscent-of-samsungs-touchwiz-notification-menu/). gottabemobile.com. . Retrieved 2011-09-07.

[274] "Email, SMS, and IM on the N900" (http://www.youtube.com/watch?v=2tktVfOmROM). youtube.com. 2009-10-21. . Retrieved 2011-09-07.

[275] "Nokia N9 Notifications" (http://www.youtube.com/watch?v=yjaONrW3znc). youtube.com. 2011-06-29. . Retrieved 2011-09-07.

[276] Palm Pre review – part 3 of 5 (http://www.youtube.com/watch?v=y-CksFO-Hv4). Youtube.com (2009-06-17). Retrieved on 2012-07-03.

[277] Nithin Ramesh (2012-06-28) Takedown_infographic (http://techie-buzz.com/mobile-news/windows-phone-wish-list-2012-2.html/print). techie-buzz.com

[278] BBE Presents: BlackBerry 6 Notifications (http://www.youtube.com/watch?v=TCNSSnrTRn0). Youtube.com (2010-12-13). Retrieved on 2012-07-03.

[279] bada 2.0 on Samsung Wave S8500 – S8500XXKK7 – notifications bar, RAM, lock speed & browser bug (http://www.youtube.com/watch?v=2fC2sEF8n8k). Youtube.com (2011-12-03). Retrieved on 2012-07-03.

[280] Bray, Tim (2010-05-27). "Android Cloud To Device Messaging" (http://android-developers.blogspot.com/2010/05/android-cloud-to-device-messaging.html). android-developers.blogspot.com. . Retrieved 2011-09-07.

[281] "Push Notifications for Windows Phone" (http://msdn.microsoft.com/en-us/library/ff402537(v=vs.92).aspx). Microsoft. 2011-08-24. . Retrieved 2011-09-07.

[282] Epstein, Zach (2010-12-20). "Samsung Bada 2.0 to get multitasking, push notifications, a new UI and more" (http://www.bgr.com/2010/12/20/samsung-bada-2-0-to-get-multitasking-push-notifications-a-new-ui-and-more/). BGR.com. . Retrieved 2011-09-07.

[283] "Apple Push Notification Service" (http://developer.apple.com/library/ios/documentation/NetworkingInternet/Conceptual/RemoteNotificationsPG/ApplePushService/ApplePushService.html). Apple Inc.. 2011-08-09. . Retrieved 2011-09-07.

[284] How to: Send and Receive Raw Notifications for Windows Phone (http://msdn.microsoft.com/en-us/library/hh202977(v=vs.92).aspx). Msdn.microsoft.com (2012-03-22). Retrieved on 2012-07-03.

[285] How to enable the orientation lock on an Android phone (http://www.youtube.com/watch?v=xcn7KsL7Tk4). Youtube.com (2011-06-01). Retrieved on 2012-07-03.

[286] Windows Phone 7 Feedback: Screen Orientation Lock (http://www.wp7feedback.com/2011/04/user-interface.html)

[287] Separate volume control NEEDED – Microsoft Answers (http://answers.microsoft.com/en-us/winphone/forum/wp7-wptips/separate-volume-control-needed/d43dccb5-022c-4e5d-a19d-b6a23ea43ec5). Answers.microsoft.com (2011-09-06). Retrieved on 2012-07-03.

[288] How To Create Your Own Ringtone on an iPhone (http://www.youtube.com/watch?v=PsIE8UjInaA). Youtube.com (2009-06-29). Retrieved on 2012-07-03.

[289] Nagy, Anton D. (2011-04-13). "Seven Great New Features of Mango for Windows Phone 7" (http://pocketnow.com/windows-phone/seven-great-new-features-of-mango-for-windows-phone-7). pocketnow.com. . Retrieved 2011-09-07.

[290] "Maemo Custom ringtones for your contacts" (http://maemo.org/packages/view/per-contact-ringtones/). maemo.org. . Retrieved 2011-09-07.

[291] "iPhone Custom Text Message Alert Sounds Come to iOS 5" (http://osxdaily.com/2011/07/11/iphone-custom-text-message-alert-sounds-come-to-ios-5/). OS X Daily. 2011-07-11. . Retrieved 2011-09-07.

[292] "How to Set Custom (SMS) Notification Ringtone on Android Phone" (http://jaxov.com/2010/10/how-to-set-custom-sms-notification-ringtone-on-android-phone/). Jaxov.com. 2010-10-8. . Retrieved 2012-06-14.

[293] "There's a Patch For That! (Part 2)" (http://www.precentral.net/there-s-patch-that-part-2). PreCentral. 2009-11-20. . Retrieved 2011-09-07.

[294] Is there any good reason Mango didn't include custom Text and Alarm tones? – Microsoft Answers (http://answers.microsoft.com/en-us/winphone/forum/wp7-wptips/is-there-any-good-reason-mango-didnt-include/d3ee6520-9a5f-44d2-bf65-f4cd63f1e482). Answers.microsoft.com (2011-09-06). Retrieved on 2012-07-03.

[295] "How to assign a ringtone and text alert to a contact on a BlackBerry Torch 9800" (http://www.youtube.com/watch?v=evaTfnYd5H8). youtube.com. 2011-02-17. . Retrieved 2011-09-07.

[296] "iOS 5 Allows You to Tap Out Custom Vibration Patterns" (http://www.macrumors.com/2011/06/08/ios-5-allows-you-to-tap-out-custom-vibration-patterns/). macrumors.com. 2011-06-08. . Retrieved 2011-09-07.

[297] "iOS 5: How To Create A Custom Vibration Alert For Each Contact" (http://iostips.net/ios-how-create-custom-vibration-alert-for-each-contact/). iostips.net. . Retrieved 2011-09-07.

[298] "Android Issue 1027: More granular notification tone, volume, and repeat controls" (http://code.google.com/p/android/issues/detail?id=1027). Google.com. 2008-10-23. . Retrieved 2011-09-07.

[299] "iMessage" (http://www.apple.com/ios/features.html#imessage). Apple Inc.. 2011-06-11. . Retrieved 2011-09-07.

[300] "Windows Phone Mango Update Threads Brings SMS, Facebook & Messenger Together" (http://socialtimes.com/windows-phone-mango-update-threads-brings-sms-facebook-messenger-together_b63786). SocialTimes.com. 2011-05-24. . Retrieved 2012-03-02.

[301] Buchanan, Ricky (2009-10-18). "iPhone/iPod Touch Voice Commands Cheat Sheet" (http://atmac.org/iphone-voice-commands-cheat-sheet). atmac.org. . Retrieved 2011-09-07.

[302] "Android Issue 11062: Offline voice commands for opening apps, controlling music player, and making calls" (http://code.google.com/p/android/issues/detail?id=11062). Google.com. 2010-09-06. . Retrieved 2011-09-07.

[303] Voice commands in Mango – Microsoft Answers (http://answers.microsoft.com/en-us/winphone/forum/wp7-wptips/voice-commands-in-mango/42abbb13-7c36-473d-bacb-4bcbdf038082?msgId=ba7dfef7-7ce7-49ce-8b58-b9e8188636c4). Answers.microsoft.com (2011-09-06). Retrieved on 2012-07-03.

[304] "iOS 5's voice recognition app will be called "Assistant" – Rumor" (http://www.edibleapple.com/ios-5s-voice-recognition-app-will-be-called-assistant-rumor/). edibleapple.com. 2011-07-25. . Retrieved 2011-09-07.

[305] Google provides a cloud based closed source implementation (Voice Actions for Android (http://www.google.com/mobile/voice-actions/index.html)) that will only work when online

[306] "Microsoft Tellme" (http://www.microsoft.com/en-us/tellme/). Microsoft. . Retrieved 2011-09-07.

[307] "Voice Control Speech Interface in Windows Phone (Mango)" (http://mynokiablog.com/2011/07/27/video-voice-control-speech-interface-in-windows-phone/). mynokiablog.com. 2011-07-27. . Retrieved 2011-09-07.

[308] Siri (software): Will Siri work fully offline on the iPhone 4S? – Quora (http://www.quora.com/Siri-software/Will-Siri-work-fully-offline-on-the-iPhone-4S). Quora.com. Retrieved on 2012-07-03.

[309] Android 4.1 Jelly Bean adds Offline Voice Typing – SlashGear (http://www.slashgear.com/android-4-1-jelly-bean-adds-offline-voice-typing-27235871/). Slashgear.com. Retrieved on 2012-07-03.

[310] "Android Issue 14462: Sound Recorder does not work on background" (http://code.google.com/p/android/issues/detail?id=14462). Google.com. 2011-02-01. . Retrieved 2011-09-07.

[311] "Maemo Recorder" (http://maemo.org/downloads/product/Maemo5/recorder/). maemo.org. . Retrieved 2011-09-07.

[312] "Android Issue 2117: Enhancement: Call Recorder" (http://code.google.com/p/android/issues/detail?id=2117). Google.com. 2009-03-02. . Retrieved 2011-09-07.

[313] "Record Your Calls With Ease With Call Recording Control" (http://www.xda-developers.com/android/record-your-calls-with-ease-with-call-recording-control/). xda-developers. 2011-02-24. . Retrieved 2011-09-07.

[314] "Two-Way In Call Recording on Windows Mobile" (http://www.brighthub.com/mobile/windows-mobile-platform/articles/30136.aspx). brighthub.com. 2011-01-13. . Retrieved 2011-09-07.

[315] "BlackBerry Forums at CrackBerry.com > Does a Call Recorder exist?" (http://forums.crackberry.com/blackberry-apps-f35/does-call-recorder-exist-4324/). CrackBerry.com. . Retrieved 2011-09-07.

[316] Ultimate Voice Recorder for S60 and Symbian ^3 by Ultimate Portable :: Symbian software @ My-Symbian.com (http://my-symbian.com/s60v3/software/applications.php?faq=1&fldAuto=214)

[317] "Ice Cream Sandwich Android Gallery app and Camera app updated" (http://www.slashgear.com/ice-cream-sandwich-android-gallery-app-and-camera-app-updated-18189047/). Slashgear. October 19, 2011. . Retrieved October 19, 2011.

[318] "10 Awesome New Features In Windows Phone 7.5 Mango" (http://www.addictivetips.com/mobile/10-awesome-new-features-in-windows-phone-7-5-mango/). Addictivetips.com. 2011-09-29. . Retrieved 2012-03-02.

[319] Nokia Europe – Nokia N9 touch screen smartphone – Specifications (http://europe.nokia.com/find-products/devices/nokia-n9/specifications). Europe.nokia.com (2011-07-01). Retrieved on 2012-07-03.

[320] "Apple Support Communities > iPhone > Using iPhone > Discussions > Picture date/time stamp" (https://discussions.apple.com/thread/1061722). Apple Inc.. 2007-07-29. . Retrieved 2011-09-07.

[321] "Android Exif Droid" (http://www.androidzoom.com/android_applications/media_and_video/exif-droid_yhg.html). androidzoom.com. 2011-01-12. . Retrieved 2011-09-07.

[322] "New iOS 5 Albums and Camera Photos Emerge" (http://www.iphonedownloadblog.com/2011/06/06/breaking-new-ios-5-albums-and-camera-photos-emerge/). iphonedownloadblog.com. 2011-06-06. . Retrieved 2011-09-07.

[323] "Android Help Forum : How do I organize photos in the Gallery?" (http://www.google.com/support/forum/p/android/thread?tid=344f64e53e51d6fe&hl=en). Google.com. 2010-02-03. . Retrieved 2011-09-07.

[324] "Microsoft Windows Phone – Add or delete albums" (http://www.microsoft.com/windowsphone/en-us/howto/wp7/pictures/add-or-delete-albums.aspx). Microsoft. 2011-06-09. . Retrieved 2011-09-07.

[325] "How to Organize Photos on N97 & Mini with Tags" (http://www.symbiandreams.com/nokia-phones/how-to-organize-photos-on-n97-mini-with-tags/). symbiandreams.com. 2010-02-28. . Retrieved 2011-09-07.

[326] Android Issue 20661: Add a descriptive text to created videos, photos and audio and search for it (http://code.google.com/p/android/issues/detail?id=20661). Code.google.com (2011-10-09). Retrieved on 2012-07-03.

[327] "Apple Support Communities > iPhone > iPhone Hardware > Discussions > Shutter noise" (https://discussions.apple.com/thread/2496491). Apple Inc.. 2010-07-07. . Retrieved 2011-09-07.

[328] "Android Issue 9854: Cannot disable shutter sound" (http://code.google.com/p/android/issues/detail?id=9854). Google.com. 2010-07-17. . Retrieved 2011-09-07.

[329] Nagy, Anton D. (2011-05-15). "WP7 Mango To Support SmartDJ, Disabling Camera Shutter Sound" (http://pocketnow.com/windows-phone/wp7-mango-to-support-smartdj-disabling-camera-sound). pocketnow.com. . Retrieved 2011-09-07.

[330] "iOS 5 Features Photo" (http://www.apple.com/ios/features.html#photos). Apple Inc.. 2011-06-11. . Retrieved 2011-09-07.

[331] Crop (http://www.windowsphone.com/en-US/apps/15cb47b7-2ef8-df11-9264-00237de2db9e). windowsphone.com

[332] Kodierer [Coder]: Pictures Lab for Windows Phone – Rene Schulte – Coding .Net and Silverlight with C# (http://kodierer.blogspot.com/2010/10/announcing-pictures-lab-for-windows.html). Kodierer.blogspot.com (2010-10-18). Retrieved on 2012-07-03.

[333] Gilbert, Jason (October 19, 2011). "Android 4.0 'Ice Cream Sandwich' Features Announced By Google" (http://www.huffingtonpost.com/2011/10/18/android-4-ice-cream-sandwich-features_n_1018881.html). Huffington Post. . Retrieved October 19, 2011.

[334] Red Eye Fix (http://www.windowsphone.com/en-US/apps/197b5b69-dc00-e011-9264-00237de2db9e). windowsphone.com

[335] Nokia N8 Symbian Anna Update: The Best Camera Phone Just Got Better (http://www.techmaish.com/nokia-n8-symbian-anna-update-best-camera-phone/). Techmaish.com (2011-09-04). Retrieved on 2012-07-03.

[336] "First look: Taking HDR photos with Apple's iOS 4.1" (http://www.appleinsider.com/articles/10/09/02/first_look_taking_hdr_photos_with_apples_ios_4_1.html). AppleInsider.com. 2010-09-02. . Retrieved 2011-09-07.

[337] "Android Issue 10949: HDR Support" (http://code.google.com/p/android/issues/detail?id=10949). Google.com. 2010-09-02. . Retrieved 2011-09-07.

[338] Whitwam, Ryan. (2010-09-03) How To Take and Create HDR Photos on Android (http://www.tested.com/news/how-to-take-and-create-hdr-photos-on-android/871/). Tested.com. Retrieved on 2012-07-03.

[339] Jeremy Horwitz (April 26, 2010). "Apple iPad Camera Connection Kit" (http://www.ilounge.com/index.php/reviews/entry/apple-ipad-camera-connection-kit/). iLounge. . Retrieved June 19, 2010.

[340] "Android Movie Studio coming to Honeycomb" (http://www.androidcentral.com/android-movie-studio-coming-honeycomb). androidcentral.com. 2011-02-07. . Retrieved 2012-06-14.

[341] "Dance Dance Revolution HD Shot and Edited on Nokia N8, showcasing camera capabilities and Video Editing features" (http://symbianworld.org/6662-dance-dance-revolution-hd-shot-and-edited-on-nokia-n8-showcasing-camera-capabilities-and-video-editing-features/). symbianworld.org. 2010-10-03. . Retrieved 2011-09-07.
[342] "Android Issue 14553: Basic sound editor (at least time trim and split)" (http://code.google.com/p/android/issues/detail?id=14553). Google.com. 2011-02-05. . Retrieved 2011-09-07.
[343] "Ringtone Maker – Apps op Android Market" (https://market.android.com/details?id=com.herman.ringtone). Market.android.com. . Retrieved 2012-03-02.
[344] "Application:FreeTether" (http://www.webos-internals.org/wiki/Application:FreeTether). WebOS Internals. 2011-04-11. . Retrieved 2011-09-07.
[345] "WiFi hotspot to be available in Windows Phone Mango, called Internet Sharing" (http://www.phonearena.com/news/WiFi-hotspot-to-be-available-in-Windows-Phone-Mango-called-Internet-Sharing_id21769). Phonearena.com. . Retrieved 2012-03-02.
[346] "Nokia N8 vs iPhone 4 – Ten Things Why Nokia N8 Is Better" (http://symbianworld.org/5003-nokia-n8-vs-iphone-4âten-things-why-nokia-n8-is-better/). symbianworld.org. 2010-06-30. . Retrieved 2011-09-07.
[347] "Android Issue 738: I hope Android will implement and support the USB host feature" (http://code.google.com/p/android/issues/detail?id=738). Google.com. 2008-05-30. . Retrieved 2011-09-07.
[348] "USB Host – Android Developers" (http://developer.android.com/guide/topics/usb/host.html). developer.android.com. 2011-09-22. . Retrieved 2011-09-28.
[349] "Nokia N8's USB On-The-Go support demoed, lesser phones turned into slaves" (http://www.engadget.com/2010/05/30/nokia-n8s-usb-on-the-go-support-demoed-lesser-phones-turned-in/). Engadget. 2010-05-30. . Retrieved 2011-09-07.
[350] Storage Options | Android Developers (http://developer.android.com/guide/topics/data/data-storage.html)
[351] Windows Phone 8 will support removable microSD storage | The Verge (http://www.theverge.com/2012/6/20/3101563/windows-phone-8-removable-micro-sd-support)
[352] Michaluk, Kevin (2010-03-18). "How to Install a Media Card into your BlackBerry" (http://crackberry.com/blackberry-101-lecture-6-how-install-microsdhc-card-your-blackberry). Crackberry.com. . Retrieved 2011-09-07.
[353] bada File System (http://developer.bada.com/help/topic/com.osp.documentation.help/html/bada_overview/bada_file_system.htm)
[354] "Pages 1.4 for iPad, iPhone, and iPod touch" (http://www.apple.com/iphone/features/pages.html). Apple Inc.. 2011-05-31. . Retrieved 2011-09-07.
[355] "Numbers 1.4 for iPad, iPhone, and iPod touch" (http://www.apple.com/iphone/features/numbers.html). Apple Inc.. 2011-05-31. . Retrieved 2011-09-07.
[356] "Keynote 1.4 for iPad, iPhone, and iPod touch" (http://www.apple.com/iphone/features/keynote.html). Apple Inc.. 2011-05-31. . Retrieved 2011-09-07.
[357] "eZReader DOC PDF Lite" (http://store.ovi.com/content/143002). . Retrieved 2012-08-02.
[358] "iPhone 4 – AirPrint" (http://www.apple.com/iphone/features/airprint.html). Apple Inc.. 2010-11-22. . Retrieved 2011-09-07.
[359] "Android Issue 1148: RFE: CUPS support for printers" (http://code.google.com/p/android/issues/detail?id=1148). Google.com. 2008-10-30. . Retrieved 2011-09-07.
[360] "Cloud Print – Apps op Android Market" (https://market.android.com/details?id=com.pauloslf.cloudprint). Market.android.com. . Retrieved 2012-03-02.
[361] "Tested: HP TouchPad Printing" (http://www.pcworld.com/article/235444/tested_hp_touchpad_printing.html). PC World. 2008-10-30. . Retrieved 2011-07-11.
[362] How do I print from a Windows 7 Phone? – Microsoft Answers (http://answers.microsoft.com/en-us/winphone/forum/wp7-wpdevices/how-do-i-print-from-a-windows-7-phone/47baecba-8c22-43d5-9d37-e3108dda223c). Answers.microsoft.com (2011-09-06). Retrieved on 2012-07-03.
[363] "Issue 9308 – android – eaac+ and aac+ decoding sound quality problems (Android 2.2) – Android – An Open Handset Alliance Project – Google Project Hosting" (http://code.google.com/p/android/issues/detail?id=9308). Code.google.com. 2010-06-24. . Retrieved 2012-03-02.
[364] "Android Core Media Formats" (http://developer.android.com/guide/appendix/media-formats.html#core). developer.android.com. 2011-09-02. . Retrieved 2011-09-07.
[365] "Media: Handling Images, Audio, Video, and Camera" (http://developer.bada.com/help/index.jsp?topic=/com.osp.devguide.help/html/user_interactive_features/media/players.htm). developer.bada.com. . Retrieved 2011-09-07.
[366] "Apple Digital AV Adapter" (http://store.apple.com/us/product/MC953ZM/A?fnode=MTc0MjU4NjE&mco=MjE1Mjg1OTQ). Apple Inc.. 2011-03-11. . Retrieved 2011-09-07.
[367] "Apple VGA Adapter" (http://store.apple.com/us/product/MC552). Apple Inc.. 2011-04-03. . Retrieved 2011-09-07.
[368] "Apple Component AV Cable" (http://store.apple.com/us/product/MC917ZM/A?mco=MTY3ODQ5OTY). Apple Inc.. . Retrieved 2011-09-07.
[369] "Apple Composite AV Cable" (http://store.apple.com/us/product/MC748ZM/A?mco=MTY3ODQ5OTY). Apple Inc.. . Retrieved 2011-09-07.
[370] Windows 7 phone with HDMI out, or connectivity to pico-projector? - Microsoft Answers (http://answers.microsoft.com/en-us/winphone/forum/wp7-wpdevices/windows-7-phone-with-hdmi-out-or-connectivity-to/

c8219066-e6d7-42b7-8646-19c223c4ce2b?msgId=ef61ebaf-b386-4d49-a772-52e647cb1ec0)

[371] "iTunes – AirPlay" (http://www.apple.com/itunes/airplay/). Apple Inc.. 2010-09-01. . Retrieved 2011-09-07.

[372] Android Issue 11412: Support DLNA in media player (http://code.google.com/p/android/issues/detail?id=11412). Code.google.com (2010-09-22). Retrieved on 2012-07-03.

[373] DLNA for WP7??? – Microsoft Answers (http://answers.microsoft.com/en-us/winphone/forum/wp7-sync/dlna-for-wp7/33a71b65-b00a-4b4b-8a75-8368b7269e5c). Answers.microsoft.com (2011-09-06). Retrieved on 2012-07-03.

[374] "LG Optimus 7 review" (http://www.engadget.com/2010/10/22/lg-optimus-7-review/). Engadget. 2010-10-22. . Retrieved 2011-09-07.

[375] Play media file in Samsung Wave from PC / Remote Media Server (AllShare-DLNA) (http://www.youtube.com/watch?v=NPYLbRKBIfY). Youtube.com (2011-02-18). Retrieved on 2012-07-03.

[376] Android 101: How To Create a Playlist (http://www.youtube.com/watch?v=EHdaICGESnI). Youtube.com (2011-04-08). Retrieved on 2012-07-03.

[377] Cogen, David (2011-06-16). "How To Create a Playlist in Windows Phone – TheUnlockrTheUnlockr" (http://theunlockr.com/2011/06/16/how-to-create-a-playlist-in-windows-phone/). Theunlockr.com. . Retrieved 2012-03-02.

[378] "A tour of the new Music +Video Hub" (http://windowsteamblog.com/windows_phone/b/windowsphone/archive/2011/06/03/a-tour-of-the-new-music-video-hub.aspx). Microsoft. 2011-06-03. . Retrieved 2011-09-07.

[379] "iOS has fine scrubbing" (http://www.youtube.com/watch?v=klRDYXutEWA#t=03m11s). youtube.com. 2010-12-11. . Retrieved 2011-09-07.

[380] "Android Issue 15505: Media player fine scrubbing (swipe to seek)" (http://code.google.com/p/android/issues/detail?id=15505). Google.com. 2011-03-15. . Retrieved 2011-09-07.

[381] "Meridian Player" (http://www.androidtapp.com/meridian-player/). androidtapp.com. 2009-05-19. . Retrieved 2011-09-07.

[382] "Listen to podcasts in half the time" (http://www.nevsblog.com/2006/06/23/listen-to-podcasts-in-half-the-time/). Nev's blog. 2006-06-23. . Retrieved 2011-09-07.

[383] "Android Issue 1961: MediaPlayer Adjustable SampleRate/Playback Speed support" (http://code.google.com/p/android/issues/detail?id=1961). Google.com. 2009-02-10. . Retrieved 2011-09-07.

[384] "Astro Player" (http://www.astroplayer.com/). Astro Player. . Retrieved 2011-09-07.

[385] Turn-by-turn navigation and Flyover in iOS 6 Maps is limited to iPhone 4S, iPad 2 or later (http://www.idownloadblog.com/2012/06/12/turn-by-turn-ios-6-iphone-4s-ipad-2/). Idownloadblog.com (2012-06-12). Retrieved on 2012-07-03.

[386] Android 2.0 gets Google Maps Navigation Beta [Video (http://androidcommunity.com/android-2-0-gets-google-maps-navigation-beta-video-20091028/) | Android Community]

[387] Windows Phone 8: the new features - 8. Offline turn-by-turn navigation, maps powered by Nokia (http://www.phonearena.com/news/Windows-Phone-8-the-new-features_id31466#8-Offline-turn-by-turn-navigation,-maps-powered-by-Nokia)

[388] "Turn-by-Turn directions, Voice to text, Bing Audio/Vision all coming to WP7 Mango" (http://www.wpcentral.com/turn-by-turn-directions-voice-text-bing-audio-vision-all-coming-wp7-mango). wpcentral. 2011-05-08. . Retrieved 2011-09-07.

[389] Mango Bing Maps turn by turn directions will require a tap for each turn (http://wmpoweruser.com/mango-bing-maps-turn-by-turn-directions-will-require-a-tap-for-each-turn/). WMPoweruser. Retrieved on 2012-07-03.

[390] Bing Maps is a bad software in Europe, ok. But how can the maps be so old and outdated ? (http://answers.microsoft.com/en-us/winphone/forum/wp7-wpfeatures/bing-maps-is-a-bad-software-in-europe-ok-but-how/de364190-6053-46ca-8e79-bfb206479aa6). Answers.microsoft.com (2011-09-06). Retrieved on 2012-07-03.

[391] "Turn By Turn Directions Comes to Windows Phone Via A-to-B (Free)" (http://mobilitydigest.com/turn-by-turn-directions-comes-to-windows-phone-via-a-to-b-free/). Mobility Digest. 2011-02-18. . Retrieved 2011-09-07.

[392] Apps with Maps: 11 iPhone GPS apps compared (http://www.macworld.com/article/1156720/gps.html). Macworld. Retrieved on 2012-07-03.

[393] "Android Issue 4471: Downloadable maps for offline navigation/location" (http://code.google.com/p/android/issues/detail?id=4471). Google.com. 2009-11-02. . Retrieved 2011-09-07.

[394] "How to download maps data on Google Maps" (http://www.androidcentral.com/how-download-maps-data-google-maps). androidcentral.com. 2011-07-06. . Retrieved 2011-09-07.

[395] http://www.phonearena.com/news/Windows-Phone-8-the-new-features_id31466#8-Offline-turn-by-turn-navigation,-maps-powered-by-Nokia

[396] Nokia Drive gets full offline access, Maps / Transport also updated – Engadget (http://www.engadget.com/2012/03/20/nokia-drive-maps-transport-windows-phone-update-lumia/). Engadget.com. Retrieved on 2012-07-03.

[397] "New in iOS 5: Suggested Routes in Maps" (http://www.idownloadblog.com/2011/06/09/ios-5-suggested-routes-maps/). idownloadblog.com. 2011-06-09. . Retrieved 2011-09-07.

[398] "Support Google Maps Alternate Routes?" (http://androidforums.com/lg-optimus-v/507001-google-maps-alternate-routes.html). Google.com. 2011-07-01. . Retrieved 2011-09-07.

[399] Head, Chris (2010-06-25). "Geek 101: Android and iOS Multitasking Compared" (http://www.pcworld.com/article/199891/geek_101_android_and_ios_multitasking_compared.html). PCWorld.com. . Retrieved 2011-09-07.

[400] Microsoft brings true, background multitasking to Windows Phone 8 -- Engadget (http://www.engadget.com/2012/06/20/microsoft-brings-true-background-multitasking-to-windows-phone/)

[401] Windows Phone 7 Camera vs iOS vs Android on Samsung Focus, iPhone, and Samsung Captivate – YouTube (http://www.youtube.com/watch?v=5u-2MBYYsmw). Youtube.com (2011-01-18). Retrieved on 2012-07-03.

[402] By: snakechia. "A look at Mango's overhauled Live Tile system [MIX11] | wpcentral | Windows Phone News, Forums, and Reviews" (http://www.wpcentral.com/mango-s-overhauled-live-tile-system-mix112). wpcentral. . Retrieved 2012-03-02.

[403] "iOS 5 – Camera" (http://www.apple.com/ios/features.html#camera). Apple Inc.. 2011-06-11. . Retrieved 2011-09-07.

[404] Christian Zibreg (2011-06-13). "Jailbreakers crack iOS 5 Notification Center widgets" (http://9to5mac.com/2011/06/13/jailbrakers-crack-ios-5-notification-center-widgets/). 9to5mac.com. . Retrieved 2011-09-07.

[405] "QuickComposer: iOS 5 Notification Center Custom Widget Will Make It Quicker To Compose SMS, Email, Tweet & More" (http://www.iphonehacks.com/2011/06/quickcomposer-ios-5-notification-center-custom-widget-to-make-it-quicker-to-compose-sms-email-tweet-more.html). iphonehacks.com. 2011-06-22. . Retrieved 2011-09-07.

[406] "Tip: Working with Dashboard Notifications [webOS 3.0 update (http://www.precentral.net/working-dashboard-notifications-webos-3-0-update)"]. PreCentral. 2011-08-29. . Retrieved 2011-09-13.<

[407] Devin Coldewey (2010-04-08). "Finally, Bluetooth keyboard support for the iPhone" (http://www.mobilecrunch.com/2010/04/08/finally-bluetooth-keyboard-support-for-the-iphone/). Mobilecrunch.com. . Retrieved 2011-09-07.

[408] "Android Issue 1147: RFE: Support for external keyboard, mouse" (http://code.google.com/p/android/issues/detail?id=1147). Google.com. 2008-10-30. . Retrieved 2011-09-07.

[409] KeyPro – Bluetooth keyboard manager (http://www.mymobilegear.com/AndroidKBDriver.php), 12/09/09

[410] "Bluetooth Keyboards On Windows Phone 7" (http://gadgetix.com/2010/12/31/bluetooth-keyboards-on-windows-phone-7/). gadgetix.com. 2010-12-31. . Retrieved 2011-09-07.

[411] Can WP7 support a wired USB Micro keyboard? – Microsoft Answers (http://answers.microsoft.com/en-us/winphone/forum/wp7-wpdevices/can-wp7-support-a-wired-usb-micro-keyboard/d3f0e6ed-50d7-4f20-80db-1bf0b81f0d63). Answers.microsoft.com (2011-09-06). Retrieved on 2012-07-03.

[412] "Celeste: Bluetooth File Sharing for iOS" (http://getceleste.com/). CocoaNuts.. 2011-09-20. . Retrieved 2011-10-12.

[413] "Android Issue 1725: Bluetooth File Transfer" (http://code.google.com/p/android/issues/detail?id=1725). Google.com. 2009-01-06. . Retrieved 2011-09-07

[414] Samsung Galaxy S II – WiFi Direct (http://www.youtube.com/watch?v=Eb3s8MJ9hSs). Youtube.com. Retrieved on 2012-07-03.

[415] Broadcom Adds Bluetooth 3.0, Wi Fi Direct to Android (http://www.broadcom.com/docs/articles/GIGAOM_02-05-10.pdf). (PDF) . Retrieved on 2012-07-03.

[416] "Windows Phone 7 OS review: From scratch" (http://www.gsmarena.com/windows_phone_7-review-521.php). *GSM Arena*. 8 October 2010. .

[417] bada 2.0 & Ecosystem FAQ's (http://developer.samsung.com/developerBlogView.do?board_id=339). Developer.samsung.com. Retrieved on 2012-07-03.

[418] Kevin Purdy (2010-12-22). "Set Up VoIP Calling on Android 2.3 Gingerbread" (http://lifehacker.com/5716008/set-up-voip-calling-on-android-23-gingerbread). lifehacker.com. . Retrieved 2011-09-07.

[419] "webOS VoIP clients arrive on the Palm Pre" (http://www.weboshelp.net/all-webos-news-articles/768-webos-voip-clients-come-to-the-palm-pre). weboshelp.net. 2009-08-02. . Retrieved 2011-09-07.

[420] Windows Phone 8 To Get Tight Skype/VoIP Integration (http://techie-buzz.com/windows-phone/windows-phone-8-to-get-tight-skypevoip-integration.html)

[421] "The Best BlackBerry VoIP Apps & Services" (http://www.voip-sol.com/the-best-blackberry-voip-apps-services/). Voip-sol.com. 2008-06-15. . Retrieved 2011-09-07.

[422] "VoIP Support in Series 40 devices" (http://www.developer.nokia.com/Community/Wiki/VoIP_support_in_Nokia_devices#Support_in_Series_40_devices). nokia.com. . Retrieved 2012-08-02.

[423] "Internet Communications Software Update for N800: Development version" (http://rtcomm.garage.maemo.org/). Rtcomm.garage.maemo.org. . Retrieved 2011-09-07.

[424] "bada VoIP applications finally allowed" (http://www.joernesdohr.com/bada/bada-voip-applications-finally-allowed/). Joernesdohr.com. 2011-04-04. . Retrieved 2011-09-07.

[425] iSSH – Zingersoft (http://www.zinger-soft.com/iSSH_features.html). Zinger-soft.com. Retrieved on 2012-07-03.

[426] Instant Cocoa – pTerm (http://www.instantcocoa.com/products/pTerm/). Instantcocoa.com (2011-03-31). Retrieved on 2012-07-03.

[427] iOS: Setting up VPN (http://support.apple.com/kb/ht1424). Support.apple.com (2011-10-26). Retrieved on 2012-07-03.

[428] Wallen, Jack. (2011-01-21) Connect to a PPTP VPN from your Android phone (http://www.techrepublic.com/blog/smartphones/connect-to-a-pptp-vpn-from-your-android-phone/2145). TechRepublic. Retrieved on 2012-07-03.

[429] Palm Support : How can I use Virtual Private Network (VPN) software with my handheld? (http://kb.hpwebos.com/wps/portal/kb/common/article/32809_en.html). Kb.hpwebos.com (1970-01-01). Retrieved on 2012-07-03.

[430] Windows Mobile 6.5 – Virtual Private Networking (http://msdn.microsoft.com/en-us/library/aa918513.aspx). Msdn.microsoft.com. Retrieved on 2012-07-03.

[431] vpn support on windows phone 7 – Microsoft Answers (http://answers.microsoft.com/en-us/winphone/forum/wp7-wptips/vpn-support-on-windows-phone-7/bac2f010-291f-4b81-98f7-cfdace86a1a2). Answers.microsoft.com (2011-09-06). Retrieved on 2012-07-03.

[432] (Dutch) Nokia Europe – Mobile VPN (http://europe.nokia.com/support/product-support/nokia-mobile-vpn). Europe.nokia.com. Retrieved on 2012-07-03.
[433] – Brainstorm: Add PPTP (VPN) support for Connnection Manager (http://maemo.org/community/brainstorm/view/add_pptp-vpn-support_for_connnection_manager/). Maemo.org. Retrieved on 2012-07-03.
[434] Nokia Support Discussions – Discussion: How to setting VPN configuration to Nokia N9? (http://discussions.europe.nokia.com/t5/Maemo-and-MeeGo-Devices/Discussion-How-to-setting-VPN-configuration-to-Nokia-N9/td-p/1196087). Discussions.europe.nokia.com. Retrieved on 2012-07-03.
[435] Re: OpenVPN without jailbreak? (http://www.guizmovpn.com/index.php?option=com_agora&task=topic&id=112&p=1&Itemid=15#p406). Guizmovpn.com. Retrieved on 2012-07-03.
[436] "OpenVPN" (http://wiki.cyanogenmod.com/wiki/OpenVPN). CyanogenMod Wiki. 2011-08-29. . Retrieved 2011-09-07.
[437] OpenVPN for Palm Pre – WebOS Internals (http://www.webos-internals.org/wiki/OpenVPN_for_Palm_Pre). Webos-internals.org (2010-10-22). Retrieved on 2012-07-03.
[438] OpenVPN for PocketPC (http://ovpnppc.ziggurat29.com/ovpnppc-main.htm). Ovpnppc.ziggurat29.com (2007-04-01). Retrieved on 2012-07-03.
[439] HowTo: Nokia N900 (OS Maemo) OpenVPN Setup Tutorial (http://www.hideipvpn.com/2010/02/howto-nokia-n900-os-maemo-openvpn-setup-tutorial/). HideIPVPN. Retrieved on 2012-07-03.
[440] Show Package openvpn (Project home:elleo) – MeeGo Community Open Build Service (https://build.pub.meego.com/package/show?package=openvpn&project=home:elleo). Build.pub.meego.com. Retrieved on 2012-07-03.
[441] "iPhone OS Enterprise Deployment Guide" (http://manuals.info.apple.com/en_US/Enterprise_Deployment_Guide.pdf) (PDF). Apple Inc.. . Retrieved 2011-09-07.
[442] "Android Issue 1386: Feature req: support WPA2-Enterprise with EAP extensions" (http://code.google.com/p/android/issues/detail?id=1386#c328). Google.com. 2008-11-28. . Retrieved 2011-09-07.
[443] Palm webOS Security Overview for Enterprise (http://www.hpwebos.com/us/assets/pdfs/business/Palm_WhitePaper_Security.pdf). hpwebos.com
[444] "Wireless EAP-TLS Authentication on Windows Phone 7" (http://answers.microsoft.com/en-us/winphone/forum/wp7-wpdevices/wireless-eap-tls-authentication-on-windows-phone-7/80808f0f-71f3-4fb0-882e-0bc8683e3c5f). Microsoft. . Retrieved 2011-09-07.
[445] Manually removing stored Wifi networks from iPad?: Apple Support Communities (https://discussions.apple.com/thread/2738932). Discussions.apple.com (2011-01-30). Retrieved on 2012-07-03.
[446] Guilfoyle, Josh (2007-12-23). "devtcg: Android RFB / VNC implementation" (http://devtcg.blogspot.com/2007/12/android-rfb-vnc-implementation.html). devtcg.blogspot.com. . Retrieved 2011-09-07.
[447] (http://www.deepthoughtsoftware.com/RemoteWin_Guide)
[448] Screenshot#iOS
[449] "android-ice-cream-sandwich-with-native-screenshot-function" (http://www.gmanews.tv/story/235870/technology/android-ice-cream-sandwich-with-native-screenshot-function). GSMarena. October 19, 2011. . Retrieved October 19, 2011.
[450] "How to make Windows Mobile screenshots?" (http://forum.xda-developers.com/showthread.php?t=363028). xda-developers.com. 2008-01-27. . Retrieved 2011-09-07.
[451] GADGET WATCH: Windows Phone Mango (http://www.smartcompany.com.au/information-technology/20110929-gadget-watch-windows-phone-mango.html). Smartcompany.com.au (2011-09-29). Retrieved on 2012-07-03.
[452] "Take a screenshot on Windows Phone 7" (http://superuser.com/questions/216092/take-a-screenshot-on-windows-phone-7). superuser.com. 2011-07-09. . Retrieved 2011-09-07.
[453] "Home brew delivers again–screen shot functionality arrive on the HTC HD7" (http://wmpoweruser.com/home-brew-delivers-againscreen-shot-functionality-arrive-on-the-htc-hd7/). WMPoweruser. 2011-01-27. . Retrieved 2011-09-07.
[454] "Screenshot for Symbian OS" (http://www.antonypranata.com/screenshot). antonypranata.com. . Retrieved 2011-09-07.
[455] How to Screen Cast and iPad: Apple Support Communities (https://discussions.apple.com/thread/3023913). Discussions.apple.com (2011-04-27). Retrieved on 2012-07-03.
[456] androidscreencast – Desktop app to control an android device remotely – Google Project Hosting (http://code.google.com/p/androidscreencast/). Code.google.com. Retrieved on 2012-07-03.
[457] Dotsisx (2009-12-02). Developer Challenge: Screencast Recorder On The Phone | Symbian-Guru.com (http://web.archive.org/web/20091205041824/http://www.symbian-guru.com/welcome/2009/12/developer-challenge-screencast-recorder-on-the-phone.html)
[458] Stewart Mitchell, "Android lets developers work closer to the metal" (http://www.pcpro.co.uk/news/365215/android-lets-developers-work-closer-to-the-metal), *PC Pro*, 11/02/11
[459] "Windows Phone 7: The AnandTech Guide" (http://www.anandtech.com/show/2969/windows-phone-7-series-the-anandtech-guide/13). AnandTech. 2010-03-21. . Retrieved 2011-09-07.
[460] "Android SDK | Android Developers" (http://developer.android.com/sdk/index.html). developer.android.com. . Retrieved 2011-09-07.
[461] "webOS Developer Center – Downloading and Installing the 3.0.2 SDK and PDK" (https://developer.palm.com/content/resources/develop/sdk_pdk_download.html). developer.palm.com. . Retrieved 2011-09-07.
[462] "Windows Mobile 6 Professional and Standard Software Development Kits Refresh" (http://www.microsoft.com/download/en/details.aspx?id=6135). Microsoft. 2007-01-05. . Retrieved 2011-09-07.

[463] "App Hub – windows phone and xbox live indie games development" (http://create.msdn.com/en-US/). Microsoft. . Retrieved 2011-09-07.
[464] "BlackBerry – Java Application Development" (http://us.blackberry.com/developers/javaappdev/). Research In Motion. . Retrieved 2011-09-07.
[465] "Nokia Developer – Series40 SDKs" (http://www.developer.nokia.com/Develop/Series_40/Platform/). Nokia. . Retrieved 2012-08-02.
[466] "Nokia Developer – Symbian SDKs" (http://www.developer.nokia.com/Resources/Tools_and_downloads/Other/Symbian_SDKs/). Nokia. . Retrieved 2011-09-07.
[467] "Nokia Qt SDK 1.0" (http://www.developer.nokia.com/info/sw.nokia.com/id/e920da1a-5b18-42df-82c3-907413e525fb/Nokia_Qt_SDK.html). Nokia. 2010-11-03. . Retrieved 2011-09-07.
[468] "Documentation/Maemo 5 Final SDK Installation – maemo.org wiki" (http://wiki.maemo.org/Documentation/Maemo_5_Final_SDK_Installation#Installing_Maemo_5_SDK_using_GUI_Installer). maemo.org. 2011-08-04. . Retrieved 2011-09-07.
[469] "MeeGo SDK" (http://developer.meego.com/meego-sdk). developer.meego.com. 2011-05-19. . Retrieved 2011-09-07.
[470] "bada SDK" (http://developer.bada.com/devtools/sdk). developer.bada.com. 2011-08-25. . Retrieved 2011-09-07.
[471] "iOS Developer Program – Apple Developer" (http://developer.apple.com/programs/ios/). Apple Inc.. . Retrieved 2011-09-07.
[472] "Developer Registration – Android Market for Developer Help" (http://www.google.com/support/androidmarket/developer/bin/answer.py?hl=en&answer=113468). Google.com. . Retrieved 2011-09-07.
[473] "Developer Program Details – HP webOS Developer Center" (https://developer.palm.com/content/resources/distribute/developing_and_distributing_with_hp/developer_program_details.html). developer.palm.com. . Retrieved 2011-09-07.
[474] "App Hub – windows phone developer registration walkthrough" (http://create.msdn.com/en-US/home/about/developer_registration_walkthrough). Microsoft. . Retrieved 2011-09-07.
[475] "Why should you become an Nokia Publisher? | Nokia Store Publish" (http://info.publish.ovi.com/?p=195). Nokia. 2010-08-31. . Retrieved 2011-09-07.
[476] Does Microsoft's refusal to allow Windows Live Region to be changed contravene European Law? (http://wmpoweruser.com/does-microsofts-refusal-to-allow-windows-live-region-to-be-changed-contravene-european-law/), WMPoweruser. Retrieved on 2012 07-03.

External links

- Java ME (http://java.sun.com/javame/index.jsp)
- Intel Mobile Platform (http://softwarecommunity.intel.com/articles/eng/1331.htm)
- Android-based smartphone shipments leapfrog Apple's iPhone (http://www.appleinsider.com/articles/10/08/12/android_based_smartphone_shipments_leapfrog_apples_iphone.html)
- Qualcomm Uplinq Mobile OS Developer Conference (Annual) (http://www.uplinq.com/)

Nokia

Nokia Corporation

NOKIA	
Type	Public
Traded as	OMX: NOK1V [1], NYSE: NOK [2], FWB: NOA3 [3]
Industry	Telecommunications equipment Internet Computer software
Founded	Tampere, Grand Duchy of Finland (1865) incorporated in Nokia (1871)
Founder(s)	Fredrik Idestam Leo Mechelin
Headquarters	Espoo, Finland
Area served	Worldwide
Key people	Risto Siilasmaa (Chairman) Stephen Elop (President & CEO)
Products	Mobile phones Smartphones Mobile computers Networks (See products listing)
Services	Maps and navigation, music, messaging and media Software solutions (See services listing)
Revenue	▼ €38.65 billion (2011)[4]
Operating income	▼ € -1.073 billion (2011)[4]
Net income	▼ € -1.164 billion (2011)[4]
Total assets	▼ €36.20 billion (2011)[4]
Total equity	▼ €11.87 billion (2011)[4]
Employees	122,148 (2012)[5]
Divisions	Mobile Solutions Mobile Phones Markets
Subsidiaries	Nokia Siemens Networks Navteq Vertu Qt Development Frameworks
Website	Nokia.com [6]

Nokia Corporation (Finnish pronunciation: [ˈnokiɑ], English /ˈnɒkiə/) (OMX: NOK1V [1], NYSE: NOK [2], FWB: NOA3 [3]) is a Finnish multinational communications and information technology corporation headquartered in Keilaniemi, Espoo, Finland.[7] Its principal products are mobile telephones and portable IT devices. It also offers Internet services including applications, games, music, maps, media and messaging through its Ovi platform, and free-of-charge digital map information and navigation services through its wholly owned subsidiary Navteq.[8] Nokia has a joint venture with Siemens, Nokia Siemens Networks, which provides telecommunications network equipment and services.[9]

Nokia has around 122,000 employees across 120 countries, sales in more than 150 countries and annual revenues of around €38 billion.[4] As of 2012 it is the world's second-largest mobile phone maker by unit sales (after Samsung), with a global market share of 22.5% in the first quarter.[10] Nokia is a public limited-liability company listed on the Helsinki, Frankfurt, and New York stock exchanges.[11] It is the world's 143rd-largest company measured by 2011 revenues according to the *Fortune Global 500*.[12]

Nokia was the world's largest vendor of mobile phones from 1998 to 2012.[10] However, over the past five years it has suffered declining market share as a result of the growing use of smartphones, principally the Apple iPhone and devices running on Google's Android operating system. As a result, its share price has fallen from a high of US$40 in 2007 to under US$3 in 2012.[13][14] Since February 2011, Nokia has had a strategic partnership with Microsoft, as part of which all Nokia smartphones will incorporate Microsoft's Windows Phone operating system (replacing Symbian). Nokia unveiled its first Windows Phone handsets, the Lumia 710 and 800, in October 2011.[15]

History

Pre-telecommunications era

Fredrik Idestam, co-founder of Nokia.

Statesman Leo Mechelin, co-founder of Nokia.

The predecessors of the modern Nokia were the Nokia Company (Nokia Aktiebolag), Finnish Rubber Works Ltd (Suomen Gummitehdas Oy) and Finnish Cable Works Ltd (Suomen Kaapelitehdas Oy).[16]

Nokia's history started in 1865 when mining engineer Fredrik Idestam established a groundwood pulp mill on the banks of the Tammerkoski rapids in the town of Tampere, in southwestern Finland in the Russian Empire and started manufacturing paper.[17] In 1868, Idestam built a second mill near the town of Nokia, fifteen kilometres (nine miles) west of Tampere by the Nokianvirta river, which had better resources for hydropower production.[18] In 1871, Idestam, with the help of his close friend statesman Leo Mechelin, renamed and transformed his firm into a share company, thereby founding the Nokia Company, the name it is still known by today.[18]

Toward the end of the 19th century, Mechelin's wishes to expand into the electricity business were at first thwarted by Idestam's opposition. However, Idestam's retirement from the management of the company in 1896 allowed Mechelin to become the company's chairman (from 1898 until 1914) and sell most shareholders on his plans, thus realizing his vision.[18] In 1902, Nokia added electricity generation to its business activities.[17]

Industrial conglomerate

In 1898, Eduard Polón founded Finnish Rubber Works, manufacturer of galoshes and other rubber products, which later became Nokia's rubber business.[16] At the beginning of the 20th century, Finnish Rubber Works established its factories near the town of Nokia and they began using Nokia as its product brand.[19] In 1912, Arvid Wickström founded Finnish Cable Works, producer of telephone, telegraph and electrical cables and the foundation of Nokia's cable and electronics businesses.[16] At the end of the 1910s, shortly after World War I, the Nokia Company was nearing bankruptcy.[20] To ensure the continuation of electricity supply from Nokia's generators, Finnish Rubber Works acquired the business of the insolvent company.[20] In 1922, Finnish Rubber Works acquired Finnish Cable Works.[21] In 1937, Verner Weckman, a sport wrestler and Finland's first Olympic Gold medalist, became president of Finnish Cable Works, after 16 years as its technical director.[22] After World War II, Finnish Cable Works supplied cables to the Soviet Union as part of Finland's war reparations. This gave the company a good foothold for later trade.[22]

The three companies, which had been jointly owned since 1922, were merged to form a new industrial conglomerate, Nokia Corporation in 1967 and paved the way for Nokia's future as a global corporation.[23] The new company was involved in many industries, producing at one time or another paper products, car and bicycle tires, footwear (including rubber boots), communications cables, televisions and other consumer electronics, personal computers, electricity generation machinery, robotics, capacitors, military communications and equipment (such as the SANLA M/90 device and the M61 gas mask for the Finnish Army), plastics, aluminium and chemicals.[24] Each business unit had its own director who reported to the first Nokia Corporation President, Björn Westerlund. As the president of the Finnish Cable Works, he had been responsible for setting up the company's first electronics department in 1960, sowing the seeds of Nokia's future in telecommunications.[25]

Eventually, the company decided to leave consumer electronics behind in the 1990s and focused solely on the fastest growing segments in telecommunications.[26] Nokian Tyres, manufacturer of tires, split from Nokia Corporation to form its own company in 1988[27] and two years later Nokian Footwear, manufacturer of rubber boots, was founded.[19] During the rest of the 1990s, Nokia divested itself of all of its non-telecommunications businesses.[26]

Telecommunications era

The seeds of the current incarnation of Nokia were planted with the founding of the electronics section of the cable division in 1960 and the production of its first electronic device in 1962: a pulse analyzer designed for use in nuclear power plants.[25] In the 1967 fusion, that section was separated into its own division, and began manufacturing telecommunications equipment. A key CEO and subsequent Chairman of the Board was *vuorineuvos* Björn "Nalle" Westerlund (1912–2009), who founded the electronics department and let it run at a loss for 15 years.

Networking equipment

In the 1970s, Nokia became more involved in the telecommunications industry by developing the Nokia DX 200, a digital switch for telephone exchanges. The DX 200 became the workhorse of the network equipment division. Its modular and flexible architecture enabled it to be developed into various switching products.[28] In 1984, development of a version of the exchange for the Nordic Mobile Telephony network was started.[29]

For a while in the 1970s, Nokia's network equipment production was separated into *Telefenno*, a company jointly owned by the parent corporation and by a company owned by the Finnish state. In 1987, the state sold its shares to Nokia and in 1992 the name was changed to Nokia Telecommunications.

In the 1970s and 1980s, Nokia developed the Sanomalaitejärjestelmä ("Message device system"), a digital, portable and encrypted text-based communications device for the Finnish Defence Forces.[30] The current main unit used by the Defence Forces is the Sanomalaite M/90 (SANLA M/90).[31]

First mobile phones

The technologies that preceded modern cellular mobile telephony systems were the various "0G" pre-cellular mobile radio telephony standards. Nokia had been producing commercial and some military mobile radio communications technology since the 1960s, although this part of the company was sold some time before the later company rationalization. Since 1964, Nokia had developed VHF radio simultaneously with Salora Oy. In 1966, Nokia and Salora started developing the ARP standard (which stands for Autoradiopuhelin, or *car radio phone* in English), a car-based mobile radio telephony system and the first commercially operated public mobile phone network in Finland. It went online in 1971 and offered 100% coverage in 1978.[34]

The Mobira Cityman 150, Nokia's NMT-900 mobile phone from 1989 (left), compared to the Nokia 1100 from 2003.[32] The Mobira Cityman line was launched in 1987.[33]

In 1979, the merger of Nokia and Salora resulted in the establishment of Mobira Oy. Mobira began developing mobile phones for the NMT (Nordic Mobile Telephony) network standard, the first-generation, first fully automatic cellular phone system that went online in 1981.[35] In 1982, Mobira introduced its first car phone, the Mobira Senator for NMT-450 networks.[35]

Nokia bought Salora Oy in 1984 and now owning 100% of the company, changed the company's telecommunications branch name to Nokia-Mobira Oy. The Mobira Talkman, launched in 1984, was one of the world's first transportable phones. In 1987, Nokia introduced one of the world's first handheld phones, the Mobira Cityman 900 for NMT-900 networks (which, compared to NMT-450, offered a better signal, yet a shorter roam). While the Mobira Senator of 1982 had weighed 9.8 kg (**unknown operator: u'strong'** lb) and the Talkman just under 5 kg (**unknown operator: u'strong'** lb), the Mobira Cityman weighed only 800 g (**unknown operator: u'strong'** oz) with the battery and had a price tag of 24,000 Finnish marks (approximately €4,560).[33] Despite the high price, the first phones were almost snatched from the sales assistants' hands. Initially, the mobile phone was a "yuppie" product and a status symbol.[24]

Nokia's mobile phones got a big publicity boost in 1987, when Soviet leader Mikhail Gorbachev was pictured using a Mobira Cityman to make a call from Helsinki to his communications minister in Moscow. This led to the phone's nickname of the "Gorba".[33]

In 1988, Jorma Nieminen, resigning from the post of CEO of the mobile phone unit, along with two other employees from the unit, started a notable mobile phone company of their own, Benefon Oy (since renamed to GeoSentric).[36] One year later, Nokia-Mobira Oy became Nokia Mobile Phones.

Involvement in GSM

Nokia was one of the key developers of GSM (Global System for Mobile Communications),[37] the second-generation mobile technology which could carry data as well as voice traffic. NMT (Nordic Mobile Telephony), the world's first mobile telephony standard that enabled international roaming, provided valuable experience for Nokia for its close participation in developing GSM, which was adopted in 1987 as the new European standard for digital mobile technology.[38][39]

Nokia delivered its first GSM network to the Finnish operator Radiolinja in 1989.[40] The world's first commercial GSM call was made on 1 July 1991 in Helsinki, Finland over a Nokia-supplied network, by then Prime Minister of Finland Harri Holkeri, using a prototype Nokia GSM phone.[40] In 1992, the first GSM phone, the Nokia 1011, was launched.[40][41] The model number refers to its launch date, 10 November.[41] The Nokia 1011 did not yet employ Nokia's characteristic ringtone, the Nokia tune. It was introduced as a ringtone in 1994 with the Nokia 2100 series.[42]

GSM's high-quality voice calls, easy international roaming and support for new services like text messaging (SMS) laid the foundations for a worldwide boom in mobile phone use.[40] GSM came to dominate the world of mobile telephony in the 1990s, in mid-2008 accounting for about three billion mobile telephone subscribers in the world, with more than 700 mobile operators across 218 countries and territories. New connections are added at the rate of 15 per second, or 1.3 million per day.[43]

Personal computers and IT equipment

In the 1980s, Nokia's computer division Nokia Data produced a series of personal computers called MikroMikko.[44] MikroMikko was Nokia Data's attempt to enter the business computer market. The first model in the line, MikroMikko 1, was released on 29 September 1981,[45] around the same time as the first IBM PC. However, the personal computer division was sold to the British ICL (International Computers Limited) in 1991, which later became part of Fujitsu.[46] MikroMikko remained a trademark of ICL and later Fujitsu. Internationally the MikroMikko line was marketed by Fujitsu as the ErgoPro.

The Nokia Booklet 3G mini laptop.

Fujitsu later transferred its personal computer operations to Fujitsu Siemens Computers, which shut down its only factory in Espoo, Finland (in the Kilo district, where computers had been produced since the 1960s) at the end of March 2000,[47][48] thus ending large-scale PC manufacturing in the country. Nokia was also known for producing very high quality CRT and early TFT LCD displays for PC and larger systems application. The Nokia Display Products' branded business was sold to ViewSonic in 2000.[49] In addition to personal computers and displays, Nokia used to manufacture DSL modems and digital set-top boxes.

Nokia re-entered the PC market in August 2009 with the introduction of the Nokia Booklet 3G mini laptop.[50]

Challenges of growth

In the 1980s, during the era of its CEO Kari Kairamo, Nokia expanded into new fields, mostly by acquisitions. In the late 1980s and early 1990s, the corporation ran into serious financial problems, a major reason being its heavy losses by the television manufacturing division and businesses that were just too diverse.[51] These problems, and a suspected total burnout, probably contributed to Kairamo taking his own life in 1988. After Kairamo's death, Simo Vuorilehto became Nokia's Chairman and CEO. In 1990–1993, Finland underwent severe economic depression,[52] which also struck Nokia. Under Vuorilehto's management, Nokia was severely overhauled. The company responded by streamlining its telecommunications divisions, and by divesting itself of the television and PC divisions.[53]

The Nokia House, Nokia's head office located by the Gulf of Finland in Keilaniemi, Espoo, was constructed between 1995 and 1997. It is the workplace of more than 1,000 Nokia employees.[24]

Probably the most important strategic change in Nokia's history was made in 1992, however, when the new CEO Jorma Ollila made a crucial strategic decision to concentrate solely on telecommunications.[26] Thus, during the rest of the 1990s, the rubber, cable and consumer electronics divisions were gradually sold as Nokia continued to divest itself of all of its non-telecommunications businesses.[26]

As late as 1991, more than a quarter of Nokia's turnover still came from sales in Finland. However, after the strategic change of 1992, Nokia saw a huge increase in sales to North America, South America and Asia.[54] The exploding worldwide popularity of mobile telephones, beyond even Nokia's most optimistic predictions, caused a logistics crisis in the mid-1990s.[55] This prompted Nokia to overhaul its entire logistics operation.[56] By 1998, Nokia's focus on telecommunications and its early investment in GSM technologies had made the company the world's largest mobile phone manufacturer,[54] a position it would hold for the next 14 consecutive years until 2012. Between 1996 and 2001, Nokia's turnover increased almost fivefold from 6.5 billion euros to 31 billion euros.[54] Logistics continues to be one of Nokia's major advantages over its rivals, along with greater economies of scale.[57][58]

Recent history

Product releases

Nokia launched its Nokia 1100 handset in 2003,[32] with over 200 million units shipped, was the best-selling mobile phone of all time and the world's top-selling consumer electronics product.[59] Also that year, Nokia is cited in Michael Saylor's 2012 book, *The Mobile Wave: How Mobile Intelligence Will Change Everything*, as one of the first players in the mobile space to recognize that there was a market opportunity in combining a game console and a mobile phone (both of which many gamers were carrying in 2003) into the N-Gage. The N-Gage was a mobile phone and game console meant to lure gamers away from the Game Boy Advance, though it cost twice as much and was said to resemble a taco.[60]

Reduction in size of Nokia mobile phones

In May 2007, Nokia released its first touch screen phone, the Nokia 7710, which was also a huge success. In November 2007, Nokia announced and released the Nokia N82, its first Nseries phone with Xenon flash. At the Nokia World conference in December 2007, Nokia announced their "Comes With Music" program: Nokia device buyers are to receive a year of complimentary access to music downloads.[61] The service became commercially available in the second half of 2008.

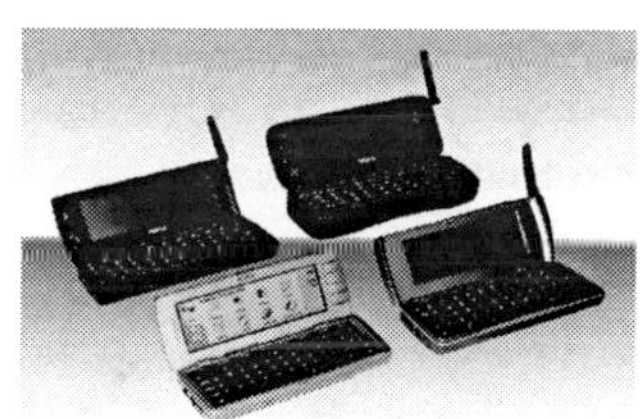

Evolution of the Nokia Communicator. Models 9000, 9110, 9210, 9300 and 9500 shown.

Nokia Productions was the first ever mobile filmmaking project directed by Spike Lee. Work began in April 2008, and the film premiered in October 2008.[62]

In 2008, Nokia released the Nokia E71 which was marketed to directly compete with the other BlackBerry-type devices offering a full "qwerty" keyboard and cheaper prices. Nokia announced in August 2009 that they will be selling a high-end Windows-based mini laptop called the Nokia Booklet 3G.[50] On 2 September 2009, Nokia launched two new music and social networking phones, the X6 and X3.[63] The Nokia X6 features 32GB of on-board memory with a 3.2" finger touch interface and comes with a music playback time of 35 hours. The Nokia X3 is a first series 40 Ovi Store-enabled device. The X3 is a music device that comes with stereo speakers, built-in FM radio, and a 3.2 megapixel camera. On 10 September 2009, Nokia unveiled the 7705 Twist, a phone sporting a square shape that swivels open to reveal a full QWERTY keypad, featuring a 3 megapixel camera, web browsing, voice commands and weighting around 3.44 ounces (**unknown operator: u'strong'** g).[64] On 9 August 2012, Nokia launched for the Indian market two new Asha range of handsets equipped with cloud accelerated Nokia browser, helping users browse the Internet faster and lower their spend on data charges. [65]

Symbian

Originally Nokia phones had a custom Nokia OS operating system developed specifically for Nokia mobile phones.

The first Nseries device, the N90, utilised the older Symbian OS 8.1 mobile operating system, as did the N70. Subsequently Nokia switched to using SymbianOS 9 for all later Nseries devices (except the N72, which was based on the N70). Newer Nseries devices incorporate newer revisions of SymbianOS 9 that include Feature Packs. The N800, N810, N900, N9 and N950 are as of April 2012 the only Nseries devices (therefore excluding Lumia devices) to not use Symbian OS. They use the Linux-based Maemo.[66]

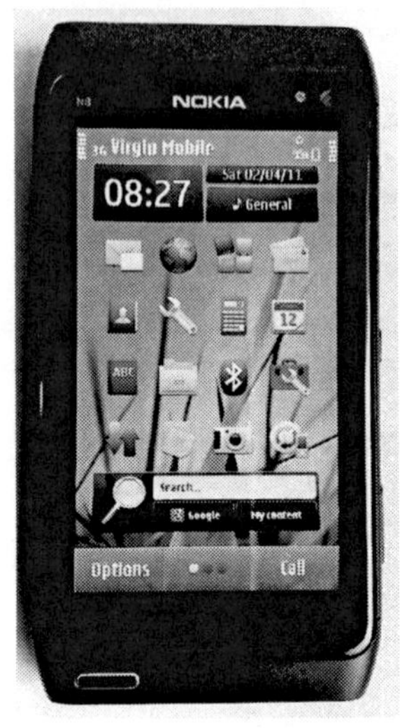

Nokia N8 running on Symbian^3 OS was Nokia's first camera phone to feature a 12 megapixel camera with Carl Zeiss Optics

Nokia stated that Maemo would be developed alongside Symbian. Maemo had since (Maemo "6" and beyond) merged with Intel's Moblin, and became MeeGo. MeeGo was later canceled and a development is now continued under name Tizen.

The Nokia N8 is the first device to function on the Symbian^3 mobile operating system.

Nokia revealed that the N8 will be the last device in its flagship N-series devices to ship with Symbian OS.[67][68]

Instead, Nokia will use Microsoft Windows Phone for its high-end flagship Lumia devices, and revealed the Nokia N9 will function on the MeeGo mobile operating system.

Alliance with Microsoft

On 11 February 2011, Nokia's CEO Stephen Elop, a former Microsoft employee, unveiled a new strategic alliance with Microsoft, and announced it would replace Symbian and MeeGo with Microsoft's Windows Phone operating system[69][70] except for mid-to-low-end devices, which would continue to run under Symbian. Nokia was also to invest into the Series 40 platform and release a single MeeGo product in 2011.[71]

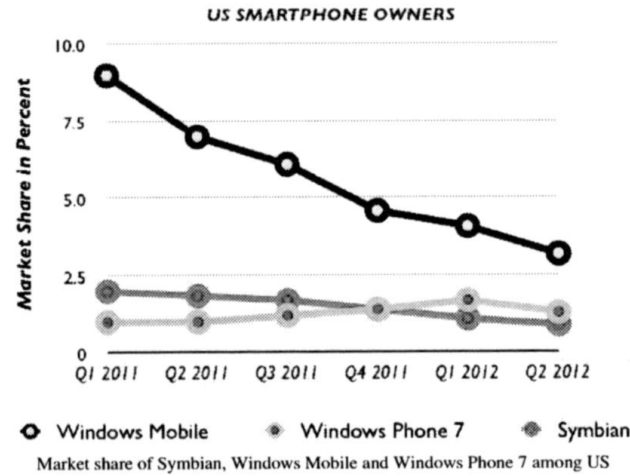

Market share of Symbian, Windows Mobile and Windows Phone 7 among US smartphone owners from Q1 2011 to Q2 2012 according to Nielsen Company.

As part of the restructuring plan, Nokia planned to reduce spending on research and development, instead customising and enhancing the software line for Windows Phone 7.[72] Nokia's "applications and content store" (Ovi) becomes integrated into the Windows Phone Marketplace, and Nokia Maps is at the heart of Microsoft's Bing and AdCenter. Microsoft provides developer tools to Nokia to replace the Qt framework, which is not supported by Windows Phone 7 devices.[73]

Symbian became described by Elop as a "franchise platform" with Nokia planning to sell 150 million Symbian devices after the alliance was set up. MeeGo emphasis was on longer-term exploration, with plans to ship "a MeeGo related product" later in 2012. Microsoft's search engine, Bing was to become the search engine for all Nokia phones. Nokia also intended to get some level of customisation on WP7.[74]

After this announcement, Nokia's share price fell about 14%, its biggest drop since July 2009.[75]

As Nokia was the largest mobile phone manufacturer worldwide at the time,[76] it was suggested the alliance would make Microsoft's Windows Phone 7 a stronger contender against Android and iOS.[73] In June 2011 Nokia was overtaken by Apple as the world's biggest smartphone maker by volume.[77] In August 2011 Chris Weber, head of Nokia's subsidiary in the U.S., stated "*The reality is if we are not successful with Windows Phone, it doesn't matter what we do (elsewhere).*" He further added "*North America is a priority for Nokia (...) because it is a key market for Microsoft.*"[78]. Additionally, while announcing an alliance with Groupon, Elop declared "The competition... is not with other device manufacturers, it's with Google."[79]

European carriers have stated that Nokia Windows phones are not good enough to compete with Apple iPhone or Samsung Galaxy phones, that "they are overpriced for what is not an innovative product" and that "No one comes into the store and asks for a Windows phone".[80]

In June 2012, Nokia chairman Risto Siilasmaa told journalists that Nokia had a back-up plan in the eventuality that Windows Phone failed to be sufficiently successful in the market.[81][82]

Plant movements

Nokia opened its Komárom, Hungary mobile phone factory on 5 May 2000.[83]

In March 2007, Nokia signed a memorandum with Cluj County Council, Romania to open a new plant near the city in Jucu commune.[84][85][86] Moving the production from the Bochum, Germany factory to a low wage country created an uproar in Germany.[87][88] Nokia recently moved its North American Headquarters to Sunnyvale.

Reorganizations

In April 2003, the troubles of the networks equipment division caused the corporation to resort to similar streamlining practices on that side, including layoffs and organizational restructuring.[89] This diminished Nokia's public image in Finland,[90][91] and produced a number of court cases and an episode of a documentary television show critical of Nokia.[92]

On February 2006, Nokia and Sanyo announced a memorandum of understanding to create a joint venture addressing the CDMA handset business. But in June, they announced ending negotiations without agreement. Nokia also stated its decision to pull out of CDMA research and development, to continue CDMA business in selected markets.[93][94][95]

In June 2006, Jorma Ollila left his position as CEO to become the chairman of Royal Dutch Shell[96] and to give way for Olli-Pekka Kallasvuo.[97][98]

In May 2008, Nokia announced on their annual stockholder meeting that they want to shift to the Internet business as a whole. Nokia no longer wants to be seen as the telephone company. Google, Apple and Microsoft are not seen as natural competition for their new image but they are considered as major important players to deal with.[99]

In November 2008, Nokia announced it was ceasing mobile phone distribution in Japan.[100] Following early December, distribution of Nokia E71 is cancelled, both from NTT docomo and SoftBank Mobile. Nokia Japan retains global research & development programs, sourcing business, and an MVNO venture of Vertu luxury phones, using docomo's telecommunications network.

In February 2012, Nokia anonunced it was laying off 4000 employees to move manufacturing from Europe and Mexico to Asia.[101]

In March 2012, Nokia annonunced it was laying off 1000 employess from its Salo, Finland factory to focus on software.[102]

Acquisitions

For a more comprehensive list, see List of acquisitions by Nokia.

On 22 September 2003, Nokia acquired Sega.com, a branch of Sega which became the major basis to develop the Nokia N-Gage device.[103]

The Nokia E55 from the business segment of the Eseries range

On 16 November 2005, Nokia and Intellisync Corporation, a provider of data and PIM synchronization software, signed a definitive agreement for Nokia to acquire Intellisync.[104] Nokia completed the acquisition on 10 February 2006.[105]

On 19 June 2006, Nokia and Siemens AG announced the companies would merge their mobile and fixed-line phone network equipment businesses to create one of the world's largest network firms, Nokia Siemens Networks.[106] Each company has a 50% stake in the infrastructure company, and it is headquartered in Espoo, Finland. The companies predicted annual sales of €16 bn and cost savings of €1.5 bn a year by 2010. About 20,000 Nokia employees were transferred to this new company.

On 8 August 2006, Nokia and Loudeye Corp. announced that they had signed an agreement for Nokia to acquire online music distributor Loudeye Corporation for approximately US $60 million.[107] The company has been developing this into an online music service in the hope of using it to generate handset sales. The service, launched on 29 August 2007, is aimed to rival iTunes. Nokia completed the acquisition on 16 October 2006.[108]

In July 2007, Nokia acquired all assets of Twango, the comprehensive media sharing solution for organizing and sharing photos, videos and other personal media.[109][110]

In September 2007, Nokia announced its intention to acquire Enpocket, a supplier of mobile advertising technology and services.[111]

In October 2007, pending shareholder and regulatory approval, Nokia bought Navteq, a U.S.-based supplier of digital mapping data, for a price of $8.1 billion.[8][112] Nokia finalized the acquisition on 10 July 2008.[113]

In September, 2008, Nokia acquired OZ Communications, a privately held company with approximately 220 employees headquartered in Montreal, Canada.[114]

On 24 July 2009, Nokia announced that it will acquire certain assets of cellity, a privately owned mobile software company which employs 14 people in Hamburg, Germany.[115] The acquisition of cellity was completed on 5 August 2009.[116]

On 11 September 2009, Nokia announced the acquisition of "certain assets of Plum Ventures, Inc, a privately held company which employed approximately 10 people with main offices in Boston, Massachusetts. Plum will complement Nokia's Social Location services".[117]

On 28 March 2010, Nokia announced the acquisition of Novarra, the mobile web browser firm from Chicago. Terms of the deal were not disclosed. Novarra is a privately held company based in Chicago, IL and provider of a mobile browser and service platform and has more than 100 employees.[118]

On 10 April 2010, Nokia announced its acquisition of MetaCarta, whose technology was planned to be used in the area of local search, particularly involving location and other services. Financial details of acquisition were not disclosed.[119]

Nokia has acquired Smarterphone in 2012.[120] Also Nokia acquired Scalado in 2012.[121]

Financial difficulties and restructuring

Amid falling sales, Nokia posted a loss of 368 million euros for Q2 2011, while in Q2 2010 had still a profit of 227 million euros. On September 2011, Nokia has announced it will lose another 3,500 jobs worldwide, including the closure of its Cluj factory in Romania.[122]

On 8 February 2012 Nokia Corp. said to cut around 4,000 jobs at smartphone manufacturing plants in Europe by the end of 2012 to move assembly closer to component supplier in Asia. It plans to cut 2,300 of the 4,400 jobs in Hungary, 700 out of 1,000 jobs in Mexico, and 1,000 out of 1,700 factory jobs in Finland.[123]

On 14 June 2012, Nokia announced to cut 10,000 jobs globally by the end of 2013[124] and shut production and research sites in Finland, Germany and Canada inline with continues loss and the stock fell to the lowest since 1996. Today, Nokia's market value is below $10 billion.[125]

In total, according to actualized and planned laid-offs Nokia will have laid off 24,500 employees by the end of 2013. Nokia has already laid off 7,000 employees in the first stage: 4,000 staff and transferred also 3,000 to services firm Accenture. Nokia also closed its factory in Cluj, Romania that decreased the workforce by 2,000 employees, and restructured the Location & Commerce business unit that decreased the workforce by 1,200 employees. In February 2012, Nokia unveiled a plan to cut 4,000 more jobs at its plants in Finland, Hungary and Mexico as it moves smartphone assembly work to Asia. The most recent plan is to cut further 10,000 jobs globally by the end of 2013.[126] Nokia had 66,267 personnel in its Devices&Services, NAVTEQ and Corporate Common Functions units combined, this has been calculated by subtracting the personnel of Nokia Siemens Networks from the total personnel of Nokia Group based on the full year report of 2010.[127] Therefore, the personnel would decrease by approximately 36 percent by the end of 2013 when compared to the end of 2010 that best depicts the lay-offs that have resulted from the strategy change in February 2011 and competition in the central mobile phone business units recently.

On 18 June 2012 Moody's downgraded Nokia rating to junk.[128] Nokia CEO admitted on 28 June 2012 that company's inability to foresee rapid changes in mobile phone industry was one of the major reasons for the problems company was facing.[129]

Corporate affairs

Corporate structure

Divisions

Since 1 July 2010, Nokia comprises three business groups: **Mobile Solutions**, **Mobile Phones** and **Markets**.[130] The three units receive operational support from the **Corporate Development Office**, led by Kai Öistämö, which is also responsible for exploring corporate strategic and future growth opportunities.[130]

On 1 April 2007, Nokia's Networks business group was combined with Siemens's carrier-related operations for fixed and mobile networks to form Nokia Siemens Networks, jointly owned by Nokia and Siemens and consolidated by Nokia.[131]

Mobile Solutions

Mobile Solutions is responsible for Nokia's portfolio of smartphones and mobile computers, including the more expensive multimedia and enterprise-class devices. The team is also responsible for a suite of internet services under the Ovi brand, with a strong focus on maps and navigation, music, messaging and media.[130] This unit is led by Anssi Vanjoki, along with Tero Ojanperä (for Services) and Alberto Torres (for MeeGo Computers).[130]

The Nokia N900, a Maemo 5 Linux based mobile Internet device and touchscreen smartphone from Nokia's Nseries portfolio.

Alberto Torres has stepped down.

Mobile Phones

The Nokia E90, a Symbian smartphone from Nokia's Eseries portfolio.

Mobile Phones is responsible for Nokia's portfolio of affordable mobile phones, as well as a range of services that people can access with them, headed by Mary T. McDowell.[130] This unit provides the general public with mobile voice and data products across a range of devices, including high-volume, consumer oriented mobile phones. The devices are based on GSM/EDGE, 3G/W-CDMA and CDMA cellular technologies.

At the end of the year 2007, Nokia managed to sell almost 440 million mobile phones which accounted for 40% of all global mobile phones sales.[132] In 2011, Nokia's market share in the mobile phone market had dropped to 27% (417 million phones).[133]

Anssi Vanjoki resigned a few days before Nokia World 2010 and under new leadership team Jo Harlow will look into the affairs of Smartphones portfolio.

On 27 April 2011, The Register reported that Nokia was secretly developing a new operating system called Meltemi aiming at the low-end market. It was believed it would be replacing the S30 and S40 operating systems. Due to low-end market customers' demand of having smartphone features in their feature phone, the OS would have included some features exclusive to high-end smartphones. On 26 July 2012, it was announced that Nokia had abandoned the Meltemi project as a cost-cutting measure.

Markets

Markets is responsible for Nokia's supply chains, sales channels, brand and marketing functions of the company, and is responsible for delivering mobile solutions and mobile phones to the market. The unit is headed by Niklas Savander.[130]

Subsidiaries

Nokia has several subsidiaries, of which the two most significant as of 2009 are Nokia Siemens Networks and Navteq.[130] Other notable subsidiaries include, but are not limited to Vertu, a British-based manufacturer and retailer of luxury mobile phones; Qt Software, a Norwegian-based software company, and OZ Communications, a consumer e-mail and instant messaging provider.

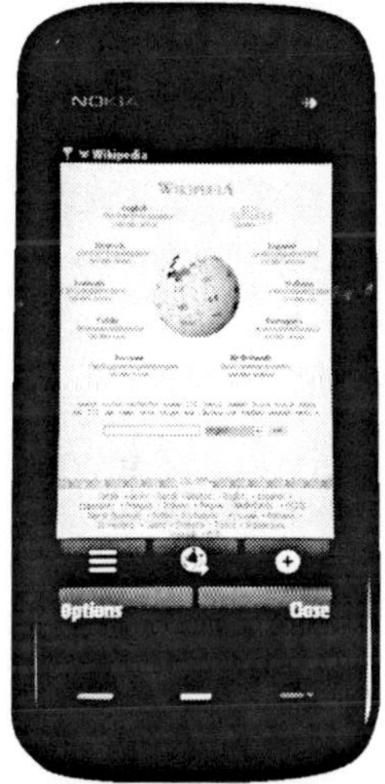

The Nokia 5800 XpressMusic, a touchscreen smartphone and portable entertainment device which emphasizes music and multimedia playback.

Until 2008 Nokia was the major shareholder in Symbian Limited, a software development and licensing company that produced Symbian OS, a smartphone operating system used by Nokia and other manufacturers. In 2008 Nokia acquired Symbian Ltd and, along with a number of other companies, created the Symbian Foundation to distribute the Symbian platform royalty free and as open source.

Nokia Siemens Networks

Nokia Siemens Networks (previously Nokia Networks) provides wireless and fixed network infrastructure, communications and networks service platforms, as well as professional services to operators and service providers.[130] Nokia Siemens Networks focuses in GSM, EDGE, 3G/W-CDMA and WiMAX radio access networks; core networks with increasing IP and multiaccess capabilities; and services.

On 19 June 2006 Nokia and Siemens AG announced the companies are to merge their mobile and fixed-line phone network equipment businesses to create one of the world's largest network firms, called Nokia Siemens Networks.[106] The Nokia Siemens Networks brand identity was subsequently launched at the 3GSM World Congress in Barcelona in February 2007.[134][135]

Navteq

Navteq is a Chicago, Illinois-based provider of digital map data and location-based content and services for automotive navigation systems, mobile navigation devices, Internet-based mapping applications, and government and business solutions.[130] Navteq was acquired by Nokia on 1 October 2007.[8] Navteq's map data is part of the Nokia Maps online service where users can download maps, use voice-guided navigation and other context-aware web services.[130] Nokia Maps is part of the Ovi brand of Nokia's Internet based online services.

Corporate governance

The control and management of Nokia is divided among the shareholders at a general meeting and the Nokia Leadership Team (left),[136] under the direction of the Board of Directors (right).[137] The Chairman and the rest of the Nokia Leadership Team members are appointed by the Board of Directors. Only the Chairman of the Nokia Leadership Team can belong to both, the Board of Directors and the Nokia Leadership Team. The Board of Directors' committees consist of the Audit Committee,[138] the Personnel Committee[139] and the Corporate Governance and Nomination Committee.[140][141]

The operations of the company are managed within the framework set by the Finnish Companies Act,[142] Nokia's Articles of Association[143] and Corporate Governance Guidelines,[144] and related Board of Directors adopted charters.

Nokia Leadership Team (as of April 2012) [136]
Stephen Elop (Chairman), b. 1963 President, CEO and Nokia Leadership Team Chairman of Nokia Corporation since 21 September 2010 Joined Nokia on 21 September 2010, Nokia Board member since 3 May 2011
Esko Aho, b. 1954 Executive Vice President, Corporate Relations and Responsibility Joined Nokia 2008, Nokia Leadership Team member since 2009 Former Prime Minister of Finland (1991–1995)
Marko Ahtisaari, b. 1969 Executive Vice President, Design Joined Nokia 2009, Nokia Leadership Team member since 1 February 2012
Jerri DeVard, b. 1958 Executive Vice President, Chief Marketing Officer Joined Nokia 2011, Nokia Leadership Team member since 1 January 2011
Colin Giles, b. 1963 Executive Vice President, Sales Joined Nokia 1992, Nokia Leadership Team member since 11 February 2011
Michael Halbherr, b. 1964 Executive Vice President, Location & Commerce Joined Nokia 2006, Nokia Leadership Team member since 1 July 2011
Jo Harlow, b. 1962 Executive Vice President, Smart Devices Joined Nokia 2003, Nokia Leadership Team member since 11 February 2011
Timo Ihamuotila, b. 1966 Executive Vice President, Chief Financial Officer With Nokia 1993–1996, rejoined 1999, Nokia Leadership Team member since 2007
Mary T. McDowell, b. 1964 Executive Vice President, Mobile Phones Joined Nokia 2004, Nokia Leadership Team member since 2004
Louise Pentland, b. 1972 Executive Vice President, Chief Legal Officer Joined Nokia 1998, Nokia Leadership Team member since 11 February 2011
Niklas Savander, b. 1962 Executive Vice President, Markets Joined Nokia 1997, Nokia Leadership Team member since 2006
Henry Tirri, b. 1956 Executive Vice President, Chief Technology Officer Joined Nokia 2004, Nokia Leadership Team member since 22 September 2011
Juha Äkräs, b. 1965 Executive Vice President, Human Resources Joined Nokia 1993, Nokia Leadership Team member since 2010
Dr. Kai Öistämö, b. 1964 Executive Vice President, Chief Development Officer Joined Nokia 1991, Nokia Leadership Team member since 2005

Board of Directors [137]
Risto Siilasmaa (Chairman), b. 1966 Board member since 2008, Chairman of the Board of Directors since 3 May 2012 Chairman of the Corporate Governance and Nomination Committee Founder and Chairman of F-Secure Corporation
Dame Marjorie Scardino (Vice Chairman), b. 1947 Board member since 2001, Vice Chairman since 2007 Member of the Corporate Governance and Nomination Committee, Member of the Personnel Committee Chief Executive Officer and member of the Board of Directors of Pearson PLC
Bruce Brown, b. 1958 Board member since 3 May 2012, Member of the Personnel Committee Chief Technology Officer of Procter & Gamble
Stephen Elop, b. 1963 Board member since 3 May 2011 President and CEO of Nokia Corporation, Chairman of the Nokia Nokia Leadership Team
Dr. Henning Kagermann, b. 1947 Board member since 2007, Chairman of the Personnel Committee, Member of the Corporate Governance and Nomination Committee Former CEO and Chairman of the Executive Board of SAP AG
Jouko Karvinen, b. 1957 Board member since 3 May 2011, Chairman of the Audit Committee, Member of the Corporate Governance and Nomination Committee CEO of Stora Enso Oyj
Helge Lund, b. 1962 Board member since 3 May 2011, Member of the Personnel Committee President and CEO of Statoil ASA
Isabel Marey-Semper, b. 1967 Board member since 2009, Member of the Audit Committee Director of Advanced Research of L'Oréal Group
Mårten Mickos, b. 1962 Board member since 3 May 2012 Chief Executive Officer of Eucalyptus Systems, Inc.
Elizabeth Nelson, b. 1960 Board member since 3 May 2012, Member of the Audit Committee Independent Corporate Advisor
Kari Stadigh, b. 1955 Board member since 3 May 2011, Member of the Personnel Committee Group CEO and President of Sampo plc

Former corporate officers

Chief Executive Officers		Chairmen of the Board of Directors [145]			
Björn Westerlund	1967–1977	Lauri J. Kivekäs	1967–1977	Simo Vuorilehto	1988–1990
Kari Kairamo	1977–1988	Björn Westerlund	1977–1979	Mika Tiivola	1990–1992
Simo Vuorilehto	1988–1992	Mika Tiivola	1979–1986	Casimir Ehrnrooth	1992–1999
Jorma Ollila	1992–2006	Kari Kairamo	1986–1988	Jorma Ollila	1999–2012
Olli-Pekka Kallasvuo	2006–2010				

International presence

In 2011 Nokia had 130,000 employees in 120 countries, sales in more than 150 countries, global annual revenue of over €38 billion, and operating loss of €1 billion.[4] It was the world's largest manufacturer of mobile phones in 2011, with global device market share of 23% in the second quarter.[76]

The Nokia Research Center, founded in 1986, is Nokia's industrial research unit consisting of about 500 researchers, engineers and scientists;[146][147] it has sites in seven countries: Finland, China, India, Kenya, Switzerland, the United Kingdom and the United States.[148] Besides its research centers, in 2001 Nokia founded (and owns) INdT – Nokia Institute of Technology, a R&D institute located in Brazil.[149] Nokia operates a total of 9 manufacturing facilities[11] located at Salo, Finland; Manaus, Brazil; Cluj, Romania; Beijing and Dongguan, China; Komárom, Hungary; Chennai, India; Reynosa, Mexico; and Changwon, South Korea.[84][150] Nokia's factory in Cluj was seized by the Romanian government in November 2011 to prevent a sale of the assets, after Nokia had accumulated a tax liability of US$ 10 million.[151] Nokia's industrial design department is headquartered in Soho in London, UK with significant satellite offices in Helsinki, Finland and Calabasas, California in the US.

Nokia is a public limited-liability company listed on the Helsinki, Frankfurt, and New York stock exchanges.[11] Nokia plays a very large role in the economy of Finland.[152][153] It is an important employer in Finland and several small companies have grown into large ones as its partners and subcontractors.[154] In 2009 Nokia contributed 1.6% to Finland's GDP, and accounted for about 16% of Finland's exports in 2006.[155]

In February 2012 Nokia announced that it was cutting 4,000 factory jobs in Finland, Hungary and Mexico (more than half of the 7,100 jobs at the three factories affected) and moving smartphone assembly to existing facilities in South Korea and China.[156]

Logos

Past

NOKIA
CONNECTING PEOPLE

Nokia introduced its "Connecting People" advertising slogan, coined by Ove Strandberg "HS Archives" (in Finnish). Helsingin Sanomat. 1 June 2003. . Retrieved 14 May 2008. and used since 1992. "NOKIA | Connecting Pople 1992 Vector Logo (AI EPS)". HDicon.com. . Retrieved 17 October 2010.This earlier version of the slogan used Times RomanTimes Roman SC (Small Caps) font.Pitkänen, Juhani (Nokia's Art Director) (3 September 2007). "Nokia Strategic Marketing, Brand Identity". Nokia Corporation. . Retrieved 14 May 2008

Present

NOKIA
Connecting People

Nokia's current logo used since 2006, "NOKIA | Connecting Pople new Vector Logo (AI EPS)". HDicon.com. . Retrieved 17 October 2010. with the redesigned "Connecting People" slogan.This slogan originally used Nokia's proprietary 'Nokia Sans' font, designed by Erik Spiekermann. "Erik Spiekermann – Furniture, Designs & Home Decor". Design Within Reach. . Retrieved 7 January 2010. This was replaced in 2011 with the 'Nokia Pure' font designed by Dalton Maag. "Our New Typeface". Nokia Little Blog of Branding. . Retrieved 3 April 2012.

Stock

Nokia, a public limited liability company, is the oldest company listed under the same name on the Helsinki Stock Exchange (since 1915).[24] Nokia's shares are also listed on the Frankfurt Stock Exchange (since 1988) and New York Stock Exchange (since 1994).[11][24]

In 2007, Nokia was valued at €110 billion; as of May 2012, it was valued at €14.8 billion. [163]

For fiscal Q2 2011 ending in June 2011, Nokia reported a net loss of €492 million, despite a €430 million payment from Apple. Nokia cited decline in its mobile phone business as the primary cause of the loss.[164]

In Q1 2012 results were bleak. Nokia lost €1.34 billion. Revenue is down almost a third from a year ago.[165] By May 2012, Nokia share price had fallen 37.5 percent since the beginning of the year, and was down 61 percent in the last year.[166][167]

Corporate culture

Nokia's official corporate culture manifesto, *The Nokia Way*, emphasises the speed and flexibility of decision-making in a flat, networked organization, although the corporation's size necessarily imposes a certain amount of bureaucracy.[168]

The Nokia House, Nokia's head office in Keilaniemi, Espoo, Finland.

The official business language of Nokia is English. All documentation is written in English, and is used in official intra-company spoken communication and e-mail.

Until May 2007, the *Nokia Values* were Customer Satisfaction, Respect, Achievement, and Renewal. In May 2007, Nokia redefined its values after initiating a series of discussions worldwide as to what the new values of the company should be. Based on the employee suggestions, the new values were defined as: Engaging You, Achieving Together, Passion for Innovation and Very Human.[168]

Online services

.mobi and the Mobile Web

Nokia was the first proponent of a Top Level Domain (TLD) specifically for the Mobile Web and, as a result, was instrumental in the launch of the .mobi domain name extension in September 2006 as an official backer.[169][170] Since then, Nokia has launched the largest mobile portal, Nokia.mobi, which receives over 100 million visits a month.[171] It followed that with the launch of a mobile Ad Service to cater to the growing demand for mobile advertisement.[172]

Ovi

Ovi, announced on 29 August 2007, is the name for Nokia's "umbrella concept" Internet services.[173] Centered on Ovi.com, it is marketed as a "personal dashboard" where users can share photos with friends, download music, maps and games directly to their phones and access third-party services like Yahoo's Flickr photo site. It has some significance in that Nokia is moving deeper into the world of Internet services, where head-on competition with Microsoft, Google and Apple is inevitable.[174]

The services offered through Ovi include the Ovi Store (Nokia's application store), the Nokia Music Store, Nokia Maps, Ovi Mail, the N-Gage mobile gaming platform available for several S60 smartphones, Ovi Share, Ovi Files, and Contacts and Calendar.[175] The Ovi Store, the Ovi application store was launched in May 2009.[176] Prior to opening the Ovi Store, Nokia integrated its software Download! store, the stripped-down MOSH repository and the widget service WidSets into it.[177]

On 23 March 2010, Nokia announced launch of its online magazine called the *Nokia Ovi*. The 44-page magazine contains articles on products by Nokia, what Ovi stands for, tips and tricks on the usage of Nokia mini laptop Booklet 3G, latest reviews of mobile applications, news about the mobile maker's services and apps such as Ovi maps, files and mail. Users can download the magazine as a PDF or view it online from the Nokia website.[178]

My Nokia

Nokia offers a free personalised service to Nokia owners called My Nokia (located at my.nokia.com).[179] Registered My Nokia users can get free services as follows:

- Tips & tricks alerts through web, e-mail and also mobile text message.
- My Nokia Backup: A free online backup service for mobile contacts, calendar logs and also various other files. This service needs GPRS connection.
- Ringtones, wallpapers, screensavers, games and other things can be downloaded free of cost.

Comes With Music

In 2007 Nokia set up their "Nokia Comes With Music" service, in partnership with Universal Music Group International, Sony BMG, Warner Music Group, EMI, and hundreds of independent labels and music aggregators, to allow 12, 18, or 24 months of unlimited free-of-charge music downloads with the purchase of a Nokia Comes With Music edition phone. Files could be downloaded on mobile devices or personal computers, and kept permanently.[61]

In January 2011 Nokia withdrew this program in 27 countries, due to its failure to gain traction with customers or mobile network operators; existing subscribers could continue to download until their contracts ended. The service continued to be offered in China, India, Indonesia, Brazil, Turkey and South Africa where take-up had been better.[180]

Nokia Messaging

On 13 August 2008 Nokia launched a beta release of "Nokia Email service", a push e-mail service, since incorporated into Nokia Messaging.[181]

Nokia Messaging operates as a centralised, hosted service that acts as a proxy between the Nokia Messaging client and the user's e-mail server. The phone does not connect directly to the e-mail server, but instead sends e-mail credentials to Nokia's servers.[182] IMAP is used as the protocol to transfer emails between the client and the server.

Controversies

NSN's provision of intercept capability to Iran

In 2008, Nokia Siemens Networks, a joint venture between Nokia and Siemens AG, reportedly provided Iran's monopoly telecom company with technology that allowed it to intercept the Internet communications of its citizens to an unprecedented degree.[183] The technology reportedly allowed it to use deep packet inspection to read and even change the content of everything from "e-mails and Internet phone calls to images and messages on social-networking sites such as Facebook and Twitter". The technology "enables authorities to not only block communication but to monitor it to gather information about individuals, as well as alter it for disinformation purposes," expert insiders told *The Wall Street Journal*. During the post-election protests in Iran in June 2009, Iran's Internet access was reported to have slowed to less than a tenth of its normal speeds, and experts suspected this was due to the use of the interception technology.[184]

The joint venture company, Nokia Siemens Networks, asserted in a press release that it provided Iran only with a 'lawful intercept capability' "solely for monitoring of local voice calls". "Nokia Siemens Networks has not provided any deep packet inspection, web censorship or Internet filtering capability to Iran," it said.[185]

In July 2009, Nokia began to experience a boycott of their products and services in Iran. The boycott was led by consumers sympathetic to the post-election protest movement and targeted at those companies deemed to be collaborating with the Islamic regime. Demand for handsets fell and users began shunning SMS messaging.[186]

Lex Nokia

In 2009, Nokia heavily supported the passing of a law in Finland that allows companies to monitor their employees' electronic communications in cases of suspected information leaking.[187] Contrary to rumors, Nokia denied that the company would have considered moving its head office out of Finland if laws on electronic surveillance were not changed.[188] The law was enacted, but with strict requirements for implementation of its provisions. As of 2010, the law has become a dead letter; no corporation has implemented it. The Finnish media dubbed the name *Lex Nokia* for this law, named after the Finnish copyright law (the so-called *Lex Karpela*) a few years back.

Nokia–Apple patent dispute

In October 2009, Nokia filed a lawsuit against Apple Inc. in the U.S. District Court of Delaware citing Apple infringed on 10 of its patents related to wireless communication including data transfer.[189] Apple was quick to respond with a countersuit filed in December 2009 accusing Nokia of 11 patent infringements. Apple's General Counsel, Bruce Sewell went a step further by stating, "Other companies must compete with us by inventing their own technologies, not just by stealing ours." This resulted in an ugly spat between the two telecom majors with Nokia filing another suit, this time with the U.S. International Trade Commission (ITC), alleging Apple of infringing its patents in "virtually all of its mobile phones, portable music players, and computers."[190] Nokia went on to ask the court to bar all U.S. imports of the Apple products including the iPhone, Mac and the iPod. Apple countersued by filing a complaint with the ITC in January 2010, the details of which are yet to be confirmed.[189]

In June 2011, Apple settled with Nokia and agreed to an estimated one time payment of $600 million and royalties to Nokia.[191] The two companies also agreed on a cross-licensing patents for some of their patented technologies.[192][193]

Environmental record

Electronic products such as cell phones impact the environment both during production and after their useful life when they are discarded and turned into electronic waste. Nokia is listed in Greenpeace's Guide to Greener Electronics that scores leading electronics manufacturers according to their policies on sustainability, climate and energy and how green their products are. In November 2011 Nokia ranked 3rd out of 15 listed electronics companies, falling two places due to its weaker performance on the Energy criteria and scoring 4.9/10.[194]

All of Nokia's mobile phones are free of toxic polyvinyl chloride (PVC) since the end of 2005 and all new models of mobile phones and accessories launched in 2010 are on track to be free of brominated compounds, chlorinated flame retardants and antimony trioxide.[194]

Nokia's voluntary take-back programme to recycle old mobile phones spans 84 countries with almost 5,000 collection points.[195] However, the recycling rate of Nokia phones was only 3–5% in 2008, according to a global consumer survey released by Nokia.[196] The majority of old mobile phones are simply lying in drawers at home and very few old devices, about 4%, are being thrown into landfill and not recycled.[196]

All of Nokia's new models of chargers meet or exceed the Energy Star requirements.[197] Nokia aims to reduce its carbon dioxide emissions by at least 18 percent in 2010 from a baseline year of 2006 and cover 50 percent of its energy needs through renewable energy sources.[198] Greenpeace is challenging the company to use its influence at the political level as number 85 on the Fortune 500 to advocate for climate legislation and call for global greenhouse gas emissions to peak by 2015.[199]

Nokia is researching the use of recycled plastics in its products, which are currently used only in packaging but not yet in mobile phones.[200]

Since 2001, Nokia has provided eco declarations of all its products and since May 2010 provides Eco profiles for all its new products.[201] In an effort to further reduce their environmental impact in the future, Nokia released a new phone concept, Remade, in February 2008.[202] The phone has been constructed of solely recyclable materials.[202] The outer part of the phone is made from recycled materials such as aluminium cans, plastic bottles, and used car tires.[203] The screen is constructed of recycled glass, and the hinges have been created from rubber tires. The interior of the phone is entirely constructed with refurbished phone parts, and there is a feature that encourages energy saving habits by reducing the backlight to the ideal level, which then allows the battery to last longer without frequent charges.

Research cooperation with universities

Nokia is actively exploring and engaging in open innovation through selective research collaborations with major universities and institutions by sharing resources and leveraging ideas. Major research collaboration is with Tampere University of Technology based in Finland. Current collaborations include:[204]

- Aalto University School of Science and Technology, Finland
- École Polytechnique Fédérale de Lausanne, Switzerland
- ETH Zurich, Switzerland
- Massachusetts Institute of Technology, United States
- Stanford University, United States
- Tampere University of Technology, Finland
- Tsinghua University, China
- University of California, Berkeley, United States
- University of Cambridge, United Kingdom
- University of Southern California, United States

Awards and recognition

The Brand Trust Report [205] published by Trust Research Advisory has ranked Nokia in the 1st position among the brands in India.

References

[1] http://www.nasdaqomxnordic.com/shares/shareinformation?Instrument=HEX24311

[2] http://www.nyse.com/about/listed/quickquote.html?ticker=nok

[3] http://www.boerse-frankfurt.de/en/equities/search/result?name_isin_wkn=NOA3

[4] "Annual Results 2011" (http://www.results.nokia.com/results/Nokia_results2011Q4e.pdf) (PDF). Nokia Corporation. 26 January 2012. . Retrieved 2 March 2012.

[5] www.results.nokia.com/results/Nokia_results2012Q1e.pdf

[6] http://www.nokia.com/

[7] "Nokia in brief (2007)" (http://www.nokia.com/NOKIA_COM_1/About_Nokia/Sidebars_new_concept/Nokia_in_brief/InBriefJuly08.pdf) (PDF). Nokia Corporation. March 2008. . Retrieved 14 May 2008.

[8] "Nokia to acquire NAVTEQ" (http://www.nokia.com/A4136001?newsid=1157198) (Press release). Nokia Corporation. 1 October 2007. . Retrieved 22 March 2009.

[9] "Company" (http://www.nokiasiemensnetworks.com/global/AboutUs/Company/?languagecode=en). Nokia Siemens Networks. . Retrieved 14 July 2009.

[10] "Samsung overtakes Nokia in mobile phone shipments" (http://www.bbc.co.uk/news/business-17865117). BBC News. 27 April 2012. . Retrieved 27 May 2012.

[11] "Nokia – FAQ" (http://www.nokia.com/about-nokia/company/faq). Nokia Corporation. . Retrieved 16 March 2009.

[12] "*Global 500* 2011" (http://money.cnn.com/magazines/fortune/global500/2011/full_list/101_200.html). Fortune. 2011. . Retrieved 12 November 2011.

[13] https://www.google.com/finance?client=ob&q=NYSE:NOK

[14] http://247wallst.com/2012/05/14/almost-nothing-can-save-nokia-except-maybe-widespread-poverty/#ixzz1v8jU0K5J

[15] "Nokia's first Windows Phone 7 handset: Is it enough?" (http://www.bbc.co.uk/news/technology-15460569). BBC News. 26 October 2011. . Retrieved 27 October 2011.

[16] "Nokia – Nokia's first century – Story of Nokia" (http://www.nokia.com/about-nokia/company/story-of-nokia/nokias-first-century). Nokia Corporation. . Retrieved 16 March 2009.

[17] "Nokia – The birth of Nokia – Nokia's first century – Story of Nokia" (http://www.nokia.com/about-nokia/company/story-of-nokia/nokias-first-century/the-birth-of-nokia). Nokia Corporation. . Retrieved 16 March 2009.

[18] Helen, Tapio. "Idestam, Fredrik (1838–1916)" (http://www.kansallisbiografia.fi/english/?id=4296). Biographical Centre of the Finnish Literature Society. . Retrieved 22 March 2009.

[19] "Nokian Footwear: History" (http://www.nokianfootwear.fi/eng/our_story/). Nokian Footwear. . Retrieved 21 March 2009.

[20] Palo-oja, Ritva; Willberg, Leena (1998) (in Finnish). *Kumi – Kumin ja Suomen kumiteollisuuden historia*. Tampere, Finland: Tampere Museums. pp. 43–53. ISBN 978-951-609-065-1.

[21] "Finnish Cable Factory – Brief History" (http://web.archive.org/web/20070705233815/http://www.kaapelitehdas.fi/php/image.php?id=4856) (PDF). Kaapelitehdas.fi (http://www.kaapelitehdas.fi). Archived from the original (http://www.kaapelitehdas.fi/php/image.php?id=4856) on 5 July 2007. . Retrieved 16 March 2009.

[22] "Nokia – Verner Weckman – Nokia's first century – Story of Nokia" (http://www.nokia.com/about-nokia/company/story-of-nokia/nokias-first-century/verner-weckman). Nokia Corporation. . Retrieved 20 March 2009.

[23] "Nokia – The merger – Nokia's first century – Story of Nokia" (http://www.nokia.com/about-nokia/company/story-of-nokia/nokias-first-century/the-merger). Nokia Corporation. . Retrieved 16 March 2009.

[24] "Nokia – Towards Telecommunications" (http://www.nokia.com/NOKIA_COM_1/About_Nokia/Sidebars_new_concept/Broschures/TowardsTelecomms.pdf) (PDF). Nokia Corporation. August 2000. . Retrieved 5 June 2008.

[25] "Nokia – First electronic dept – Nokia's first century – Story of Nokia" (http://www.nokia.com/about-nokia/company/story-of-nokia/nokias-first-century/first-electronic-dept). Nokia Corporation. . Retrieved 16 March 2009.

[26] "Nokia – Jorma Ollila – Mobile revolution – Story of Nokia" (http://www.nokia.com/about-nokia/company/story-of-nokia/mobile-revolution/jorma-ollila). Nokia Corporation. . Retrieved 21 March 2009.

[27] "History in brief" (http://www.nokiantyres.com/history-in-brief). Nokian Tyres. . Retrieved 22 March 2009.

[28] Kaituri, Tommi (2000). "Automaattisten puhelinkeskusten historia" (http://www.cs.helsinki.fi/u/kerola/tkhist/k2000/alustukset/puhelinkeskukset/) (in Finnish). . Retrieved 21 March 2009.

[29] Palmberg, Christopher; Martikainen, Olli (23 May 2003). "Overcoming a Technological Discontinuity – The Case of the Finnish Telecom Industry and the GSM" (http://www.etla.fi/files/677_dp855.pdf) (PDF). The Research Institute of the Finnish Economy. . Retrieved 14 June 2009.

[30] "Puolustusvoimat: Kalustoesittely – Sanomalaitejärjestelmä" (http://www.mil.fi/maavoimat/kalustoesittely/index.dsp?level=81) (in Finnish). The Finnish Defence Forces. 15 June 2005. . Retrieved 14 May 2008.

[31] "The Finnish Defence Forces: Presentation of equipment: Message device" (http://www.mil.fi/maavoimat/kalustoesittely/00030_en.dsp). The Finnish Defence Forces. . Retrieved 14 June 2009.

[32] "Nokia 1100 phone offers reliable and affordable mobile communications for new growth markets" (http://press.nokia.com/PR/200308/915317_5.html) (Press release). Nokia Corporation. 27 August 2003. . Retrieved 26 May 2009.

[33] "Nokia – Mobira Cityman – The move to mobile – Story of Nokia" (http://www.nokia.com/about-nokia/company/story-of-nokia/the-move-to-mobile/mobira-cityman). Nokia Corporation. . Retrieved 14 May 2008.

[34] Juutilainen, Matti. "Siirtyvä tietoliikenne, luennot 7–8: Matkapuhelinverkot" (http://www.it.lut.fi/kurssit/06-07/Ti5312600/luentokalvot/luento07-08.pdf) (in Finnish) (PDF). Lappeenranta University of Technology. . Retrieved 22 March 2009.

[35] "Nokia – Mobile era begins – The move to mobile – Story of Nokia" (http://www.nokia.com/about-nokia/company/story-of-nokia/the-move-to-mobile/mobile-era-begins). Nokia Corporation. . Retrieved 20 March 2009.

[36] Karttunen, Anu (2 May 2003). "Tähdet syöksyvät, Benefon" (http://www.talouselama.fi/sijoittaminen/article165594.ece) (in Finnish). *Talouselämä* (Talentum Oyj). . Retrieved 28 July 2009.

[37] "Nokia´s Pioneering GSM Research and Development to be Awarded by Eduard Rhein Foundation" (http://press.nokia.com/1997/10/17/nokiaÂ´s-pioneering-gsm-research-and-development-to-be-awarded-by-eduard-rhein-foundation) (Press release). Nokia Corporation. 17 October 1997. . Retrieved 7 April 2012.

[38] "Global Mobile Communication is 20 years old" (http://gsmworld.com/newsroom/press-releases/2070.htm) (Press release). GSM Association. 6 September 2007. . Retrieved 23 March 2009.

[39] "Happy 20th birthday, GSM" (http://news.zdnet.co.uk/leader/0,1000002982,39289154,00.htm). *ZDNet.co.uk* (CBS Interactive). 7 September 2007. . Retrieved 23 March 2009.

[40] "Nokia – First GSM call – The move to mobile – Story of Nokia" (http://www.nokia.com/about-nokia/company/story-of-nokia/the-move-to-mobile/first-gsm-call). Nokia Corporation. . Retrieved 20 March 2009.

[41] Smith, Tony (9 November 2007). "15 years ago: the first mass-produced GSM phone" (http://www.reghardware.co.uk/2007/11/09/ft_nokia_1011/). *Register Hardware*. Situation Publishing Ltd. . Retrieved 23 March 2009.

[42] "Nokia – Nokia Tune – Mobile revolution – Story of Nokia" (http://www.nokia.com/about-nokia/company/story-of-nokia/mobile-revolution/nokia-tune). Nokia Corporation. . Retrieved 23 March 2009.

[43] "3 Billion GSM Connections On The Mobile Planet – Reports The GSMA" (http://www.gsmworld.com/newsroom/press-releases/2008/1108.htm). GSM Association. 16 April 2008. . Retrieved 21 March 2009.

[44] "Nokia MikroMikko 1" (http://www.old-computers.com/museum/computer.asp?st=1&c=630). Old-Computers.com. . Retrieved 14 May 2008.

[45] "Net – Fujitsun asiakaslehti, Net-lehden historia: 1980-luku" (http://www.fujitsuservices.fi/historia/net/1980.htm) (in Finnish). Fujitsu Services Oy, Finland. . Retrieved 22 March 2009.

[46] "Historia: 1991–1999" (http://www.fujitsu.com/fi/about/history/1991/) (in Finnish). Fujitsu Services Oy, Finland. . Retrieved 22 March 2009.

[47] Hietanen, Juha (28 February 2000). "Closure of Fujitsu Siemens plant – a repeat of Renault Vilvoorde?" (http://www.eiro.eurofound.eu.int/2000/02/feature/fi0002136f.html). EIRO, European Industrial Relations Observatory on-line. . Retrieved 14 May 2008.

[48] Hietanen, Juha (28 February 2000). "Fujitsu Siemens tehdas suljetaan – toistuiko Renault Vilvoord?" (http://www.eurofound.europa.eu/eiro/2000/02/word/fi0002136ffi.doc) (in Finnish) (DOC). EIRO, European Industrial Relations Observatory on-line. . Retrieved 14 May 2008.

[49] "ViewSonic Corporation Acquires Nokia Display Products' Branded Business" (http://press.nokia.com/PR/200001/775025_5.html) (Press release). Nokia Corporation. 17 January 2000. . Retrieved 22 March 2009.

[50] "Nokia Booklet 3G brings all day mobility to the PC world" (http://www.nokia.com/press/press-releases/showpressrelease?newsid=1336683) (Press release). Nokia Corporation. 24 August 2009. . Retrieved 26 August 2009.

[51] Pietilä, Antti-Pekka (27 September 2000). "Kari Kairamon nousu ja tuho" (http://www.taloussanomat.fi/arkisto/2000/09/27/kari-kairamon-nousu-ja-tuho/200026243/12) (in Finnish). *Taloussanomat*. . Retrieved 21 March 2009.

[52] "Finland: How bad policies turned bad luck into a recession" (http://www.cepr.org/PRESS/EP29 finland.htm). Centre for Economic Policy Research. . Retrieved 5 April 2009.

[53] Häikiö, Martti; translated by Hackston, David (2001) (in Finnish). *Nokia Oyj:n historia 1–3 (A history of Nokia plc 1–3)* (http://www.finlit.fi/booksfromfinland/bff/102/nokia.htm). Helsinki: Edita. ISBN 951-37-3467-6. . Retrieved 21 March 2008.

[54] "Nokia – Leading the world – Mobile revolution – Story of Nokia" (http://www.nokia.com/about-nokia/company/story-of-nokia/mobile-revolution/leading-the-world). Nokia Corporation. . Retrieved 21 March 2009.

[55] Reinhardt, Andy (3 August 2006). "Nokia's Magnificent Mobile-Phone Manufacturing Machine" (http://www.businessweek.com/globalbiz/content/aug2006/gb20060803_618811.htm). *BusinessWeek Online Europe*. . Retrieved 21 March 2009.

[56] Professor Voomann, Thomas E.; Cordon, Carlos (1998). "Nokia Mobile Phones: Supply Line Management" (http://stuff.mit.edu/afs/athena/course/15/15.795/Nokia Supply Chain Case Study.pdf) (PDF). Lausanne, Switzerland: IMD – International Institute for Management Development. . Retrieved 21 March 2009.

[57] Ewing, Jack (30 July 2007). "Why Nokia Is Leaving Moto in the Dust" (http://www.businessweek.com/magazine/content/07_31/b4044050.htm). *BusinessWeek Online*. . Retrieved 21 March 2009.

[58] Lin, Porter; Khan, Raedeep; Piekute, Vaida; Luhtasela, Jussi; Fang, Debby (1 December 2005). "Supply Chain Management Case Nokia" (http://imba.nccu.edu.tw/OIP/EXchange/Docs/F04/mis/final/group6/SCM in Nokia - Written report-V1.0.pdf) (PDF). IMBA, College of Commerce, National Chengchi University. . Retrieved 21 March 2009.

[59] Virki, Tarmo (5 March 2007). "Nokia's cheap phone tops electronics chart" (http://www.reuters.com/article/technologyNews/idUSL0262945620070503). Reuters. . Retrieved 14 May 2008.

[60] Saylor, Michael (2012). *The Mobile Wave: How Mobile Intelligence Will Change Everything*. Perseus Books/Vanguard Press. p. 81.

[61] "Nokia World 2007: Nokia outlines its vision of Internet evolution and commitment to environmental sustainability" (http://www.nokia.com/A4136001?newsid=1172937) (Press release). Nokia Corporation. 4 December 2007. . Retrieved 14 May 2008.

[62] "Nokia Productions and Spike Lee premiere the world's first social film" (http://www.nokia.com/press/press-releases/showpressrelease?newsid=1259528) (Press release). Nokia Corporation. 14 October 2008. . Retrieved 12 June 2009.

[63] "Nokia seizes social internet and amplifies music experience" (http://www.nokia.com/press/press-releases/showpressrelease?newsid=1338896) (Press release). Nokia Corporation. 2 September 2009. . Retrieved 12 October 2009.

[64] "Nokia 7705 Twist launched Stateside on Verizon (photo gallery)" (http://conversations.nokia.com/2009/09/10/nokia-7705-twist-launched-stateside-on-verizon-photo-gallery/). Nokia Corporation. 10 September 2009. . Retrieved 11 October 2009.

[65] "Nokia launches two new handsets under 'Asha' range" (http://economictimes.indiatimes.com/tech/hardware/nokia-launches-two-new-handsets-under-asha-range/articleshow/15421139.cms). 09-08-2012. .

[66] "Home of the Maemo community" (http://maemo.org/). maemo.org. . Retrieved 12 November 2011.

[67] Paul, Ryan (25 June 2010). "Nokia picks MeeGo Linux, not Symbian, for flagship phones" (http://arstechnica.com/open-source/news/2010/06/nokia-to-use-meego-linux-and-not-symbian-for-flagship-phones.ars). Ars Technica. . Retrieved 12 November 2011.

[68] Sherwood, James (13 August 2009). "Nokia exec denies Symbian Maemo swap claim" (http://www.reghardware.co.uk/2009/08/13/nokia_denies_maemo/). reghardware. . Retrieved 12 November 2011.

[69] "Nokia announces strategic partnership with Microsoft, will use WP7 as primary OS" (http://www.techit.in/2011/02/nokia-announces-strategic-partnership-with-microsoft-will-use-wp7-as-primary-os/). TechIt.in. .

[70] "Missed the historic Nokia+Microsoft event today? See it here!" (http://www.techit.in/2011/02/missed-the-historic-nokiamicrosoft-event-today-see-it-here/). TechIt.in. .

[71] "Nokia and Microsoft form partnership" (http://www.bbc.co.uk/news/business-12427680). BBC. 11 February 2011. . Retrieved 12 February 2011.

[72] "RIP: Symbian" (http://www.engadget.com/2011/02/11/rip-symbian/). Engadget. .

[73] "Capitulation: Nokia adopts Windows Phone 7" (http://arstechnica.com/gadgets/news/2011/02/nokia-adopts-windows-phone-7-as-primary-platform.ars). ArsTechnica. 11 February 2011. . Retrieved 12 February 2011.

[74] "Nokia will be able to customize 'everything' in Windows Phone 7, but likely won't" (http://www.engadget.com/2011/02/11/nokia-will-be-able-to-customize-everything-in-windows-phone-7). Engadget. .

[75] ben-Aaron, Diana (11 February 2011). "Nokia Falls Most Since July 2009 After Microsoft Deal" (http://www.bloomberg.com/news/2011-02-11/nokia-joins-forces-with-microsoft-to-challenge-dominance-of-apple-google.html). Bloomberg. .

[76] "Gartner Says Sales of Mobile Devices in Second Quarter of 2011 Grew 16.5 Percent Year-on-Year; Smartphone Sales Grew 74 Percent" (http://www.gartner.com/it/page.jsp?id=1764714) (Press release). Gartner. 11 August 2011. . Retrieved 29 September 2011.

[77] Ward, Andrew (21 July 2011). "Apple overtakes Nokia in smartphone stakes" (http://www.ft.com/cms/s/0/4d7fd1e2-b38e-11e0-b56c-00144feabdc0.html#axzz1SlVal4Cs). *Financial Times*. . Retrieved 21 July 2011.

[78] Fried, Ina (9 August 2011). "Nokia to Exit Symbian, Low-End Phone Businesses in North America" (http://allthingsd.com/20110809/exclusive-nokia-to-exit-symbian-low-end-phone-businesses-in-north-america/). All Things Digital. . Retrieved 9 August 2011.

[79] "Nokia's Lumia phones to show Groupon offers on maps" (http://www.bbc.com/news/technology-19093769). BBC. .

[80] Mobile operators unconvinced by Nokia's revival bid | Reuters (http://uk.reuters.com/article/2012/04/17/uk-nokia-telcos-idUKBRE83G08Z20120417)

[81] Chris Smith (30 June 2012). "Nokia promises back-up plan if Windows Phone fails" (http://www.techradar.com/news/phone-and-communications/mobile-phones/nokia-promises-back-up-plan-if-windows-phone-fails-1087600). Techradar. .

[82] "Nokia's Siilasmaa: Goal to regain competitiveness" (http://yle.fi/uutiset/nokias_siilasmaa_goal_to_regain_competitiveness/6199219). YLS Uutiset. 28 June 2012. .

[83] "Hungarian and Finnish Prime Ministers Inaugurate Nokia's "Factory of the Future" in Komárom" (http://press.nokia.com/PR/200005/780293_5.html) (Press release). Nokia Corporation. 5 May 2000. . Retrieved 22 March 2009.

[84] "Nokia to set up a new mobile device factory in Romania" (http://www.nokia.com/A4136001?newsid=1114420) (Press release). Nokia Corporation. 26 March 2007. . Retrieved 14 May 2008.

[85] "Nokia to open cell phone plant near Cluj" (http://web.archive.org/web/20080507194303/http://www.boston.com/news/world/europe/articles/2007/03/22/nokia_to_open_cell_phone_plant_near_cluj/). Associated Press. Boston.com. 22 March 2007. Archived from the original (http://www.boston.com/news/world/europe/articles/2007/03/22/nokia_to_open_cell_phone_plant_near_cluj/) on 7 May 2008. . Retrieved 14 May 2008.

[86] "Nokia to build mobile phone plant in Romania" (http://www.hs.fi/english/article/Nokia+to+build+mobile+phone+plant+in+Romania/1135226144930). *Helsingin Sanomat*. 27 March 2007. . Retrieved 14 May 2008.

[87] "German Politicians Return Cell Phones Amid Nokia Boycott Calls" (http://www.dw-world.de/dw/article/0,2144,3076534,00.html). *Deutsche Welle*. 18 January 2008. . Retrieved 22 March 2009.

[88] "German State Demands €60 Million from Nokia" (http://www.spiegel.de/international/business/0,1518,540699,00.html). *Der Spiegel*. 11 March 2008. . Retrieved 22 March 2009.

[89] "Nokia Networks takes strong measures to reduce costs, improve profitability and strengthen leadership position" (http://press.nokia.com/PR/200304/898905_5.html) (Press release). Nokia Corporation. 10 April 2003. . Retrieved 14 May 2008.

[90] "Nokia Networks to shed 1,800 jobs worldwide; majority of impact felt in Finland" (http://www2.hs.fi/english/archive/news.asp?id=20030411IE6). *Helsingin Sanomat*. 11 April 2003. . Retrieved 14 May 2008.

[91] Leyden, John (10 April 2003). "Nokia Networks axes 1,800 staff" (http://www.theregister.co.uk/2003/04/10/nokia_networks_axes/). *The Register*. . Retrieved 14 May 2008.

[92] "Nokia's Law (transcription)" (http://www.yle.fi/mot/kj050117/englishscript.htm). YLE TV1, Mot. 17 January 2005. . Retrieved 14 May 2008.

[93] "Nokia and Sanyo proposed new company will not proceed" (http://www.nokia.com/A4136002?newsid=1059331) (Press release). Nokia Corporation. 26 June 2006. . Retrieved 14 May 2008.

[94] "Nokia decides not to go forward with Sanyo CDMA partnership and plans broad restructuring of its CDMA business" (http://www.nokia.com/A4136002?newsid=1059329) (Press release). Nokia Corporation. 22 June 2006. . Retrieved 14 May 2008.

[95] "Nokia and Sanyo Announce Intent to Form a Global CDMA Mobile Phones Business" (http://www.nokia.com/A4136002?newsid=1034612) (Press release). Nokia Corporation. 14 February 2006. . Retrieved 14 May 2008.

[96] "Shell appoints Jorma Ollila as new Chairman" (http://www.shell.com/home/content/media/news_and_library/press_releases/2005/pr_announcement_04082005.html) (Press release). Royal Dutch Shell. 4 August 2005. . Retrieved 22 March 2009.

[97] "Nokia moves forward with management succession plan" (http://www.nokia.com/A4136002?newsid=1004430) (Press release). Nokia Corporation. 1 August 2005. . Retrieved 22 March 2009.

[98] Repo, Eljas; Melender, Tommi (19 September 2005). "Changing the guard at Nokia – Olli-Pekka Kallasvuo takes the helm" (http://finland.fi/netcomm/news/showarticle.asp?intNWSAID=41296&LAN=ENG). *Ministry for Foreign Affairs of Finland*. Virtual Finland. . Retrieved 22 March 2009.

[99] Kallasvuo, Olli-Pekka; President and CEO (8 May 2008). "2008 Nokia Annual General Meeting (transcription)" (http://nds1.nokia.com/NOKIA_COM_1/Microsites/AGM_2008/pdf/OPK_AGM_2008_ENGLISH.pdf) (PDF). Helsinki Fair Centre, Amfi Hall: Nokia Corporation. . Retrieved 12 June 2009.

[100] " □□□□□□□□ " (http://www.nokia.co.jp/about/release_081127.shtml) (in Japanese). – –. 27 November 2008.. Retrieved 5 December 2008.
[101] Nokia Will Lay off 4,000 and Move More Manufacturing to Asia | PCWorld Business Center (http://www.pcworld.com/businesscenter/article/249507/nokia_will_lay_off_4000_and_move_more_manufacturing_to_asia.html)
[102] Nokia Lays Off 1,000 Employees From Finnish Plant, Will Focus On Software (http://i2mag.com/nokia-lays-off-1000-employees-from-finnish-plant-will-focus-on-software/)
[103] "Nokia completes acquisition of assets of Sega.com Inc." (http://www.nokia.com/A4136002?newsid=918198) (Press release). Nokia Corporation. 22 September 2003.. Retrieved 16 March 2009.
[104] "Nokia to extend leadership in enterprise mobility with acquisition of Intellisync" (http://www.nokia.com/A4136002?newsid=1021663) (Press release). Nokia Corporation. 16 November 2005.. Retrieved 22 March 2009.
[105] "Nokia completes acquisition of Intellisync" (http://www.nokia.com/A4136002?newsid=1034184) (Press release). Nokia Corporation. 10 February 2006.. Retrieved 22 March 2009.
[106] "Nokia and Siemens to merge their communications service provider businesses" (http://www.nokia.com/A4136002?newsid=1057716) (Press release). Nokia Corporation. 19 June 2006.. Retrieved 22 March 2009.
[107] "Nokia to acquire Loudeye and launch a comprehensive mobile music experience" (http://www.nokia.com/A4136002?newsid=1067845) (Press release). Nokia Corporation. 8 August 2006.. Retrieved 14 May 2008.
[108] "Nokia completes Loudeye acquisition" (http://www.nokia.com/A4136001?newsid=1081455) (Press release). Nokia Corporation. 16 October 2006.. Retrieved 14 May 2008.
[109] "Nokia acquires Twango to offer a comprehensive media sharing experience" (http://www.nokia.com/A4136001?newsid=1141417) (Press release). Nokia Corporation. 24 July 2007.. Retrieved 14 May 2008.
[110] "Nokia Acquires Twango – Frequently Asked Questions (FAQ)" (http://www.nokia.com/NOKIA_COM_1/Press/Materials/NokiaTwangoFAQ.pdf) (PDF). Nokia Corporation.. Retrieved 14 May 2008.
[111] "Nokia to acquire Enpocket to create a global mobile advertising leader" (http://www.nokia.com/A4136001?newsid=1153772) (Press release). Nokia Corporation. 17 September 2007.. Retrieved 14 May 2008.
[112] Niccolai, James (1 October 2007). "Nokia buys mapping service for $8.1 billion" (http://www.infoworld.com/article/07/10/01/Nokia-buys-mapping-service-for-8.1-billion_1.html). *IDG News Service* (InfoWorld).. Retrieved 14 May 2008.
[113] "Nokia completes its acquisition of NAVTEQ" (http://www.nokia.com/A4136001?newsid=1235107) (Press release). Nokia Corporation. 10 July 2008.. Retrieved 22 March 2009.
[114] "Nokia to acquire leading consumer email and instant messaging provider OZ Communications" (http://news.taume.com/World-Business/Business-Finance/Nokia-to-acquire-leading-consumer-email-and-instant-messaging-provider-OZ-Communications-6922). *Taume News*. 30 September 2008.. Retrieved 30 September 2008.
[115] "Nokia to acquire cellity" (http://www.nokia.com/press/press-releases/showpressrelease?newsid=1330831) (Press release). Nokia Corporation. 24 July 2009.. Retrieved 4 August 2009.
[116] "Nokia completes acquisition of cellity" (http://www.nokia.com/press/press-releases/showpressrelease?newsid=1332884) (Press release). Nokia Corporation. 5 August 2009.. Retrieved 6 August 2009.
[117] "Nokia has acquired Plum" (http://www.nokia.com/press/press-releases/showpressrelease?newsid=1340931) (Press release). Nokia Corporation. 11 September 2009.. Retrieved 28 January 2010.
[118] "Nokia Acquires Browser Firm Novarra" (http://techie-buzz.com/mobile-news/nokia-acquires-browser-firm-novarra.html). Techie-buzz. 28 March 2010.. Retrieved 29 March 2010.
[119] "Nokia Acquires MetaCarta" (http://www.informationweek.com/news/mobility/smart_phones/showArticle.jhtml?articleID=224202519). informationweek. 11 April 2010.. Retrieved 12 April 2010.
[120] http://www.theverge.com/mobile/2012/1/7/2690366/nokia-buys-smarterphone-developer-of-feature-phone-operating-system
[121] http://www.pocket-lint.com/news/46116/nokia-acquires-scalado-for-better-image-quality
[122] "Nokia to cut 3500 jobs worldwide; to shut Romania factory" (http://www.moneycontrol.com/news/world-news/nokia-to-cut-3500-jobs-worldwide-to-shut-romania-factory_592235.html). 29 September 2011..
[123] Moen, Arild (8 February 2012). "Nokia to Cut 4,000 Jobs" (http://online.wsj.com/article/SB10001424052970204136404577210401816583074.html). *The Wall Street Journal*..
[124] "Nokia to cut 10,000 jobs; shut units" (http://www.thehindu.com/business/companies/article3528038.ece). *The Hindu*. 14 June 2012..
[125] ben-Aaron, Diana (14 June 2012). "Nokia to Cut 10,000 Jobs as Elop Tries to Stanch Losses" (http://www.bloomberg.com/news/2012-06-14/nokia-to-cut-10-000-jobs-as-elop-tries-to-stanch-losses.html). *Bloomberg*..
[126] http://timesofindia.indiatimes.com/tech/itslideshow/14151500.cms
[127] http://i.nokia.com/blob/view/-/263824/data/1/-/Request-Nokia-in-2010-pdf.pdf
[128] Nokia Downgraded to Junk - Zacks.com (http://www.zacks.com/stock/news/77208/nokia-downgraded-to-junk)
[129] "Nokia CEO Stephen Elop admits failure to foresee fast-changing industry" (http://economictimes.indiatimes.com/news/international-business/nokia-ceo-stephen-elop-admits-failure-to-foresee-fast-changing-industry/articleshow/14466105.cms). 28 June 2012..
[130] "Structure" (http://www.nokia.com/about-nokia/company/structure). Nokia Corporation. 1 October 2009.. Retrieved 28 December 2009.
[131] "Nokia Siemens Networks starts operations and assumes a leading position in the communications industry" (http://www.nokia.com/A4136002?newsid=1116423) (Press release). Nokia Corporation. 2 April 2007.. Retrieved 7 April 2009.

[132] "Nokia's 25 percent profit jump falls short of expectations" (http://www.usatoday.com/money/economy/2008-04-17-173945271_x.htm). Associated Press. USA Today. 17 April 2008. . Retrieved 14 May 2008.

[133] "Worldwide Mobile Phone Market Maintains Its Growth Trajectory in the Fourth Quarter Despite Soft Demand for Feature Phones, According to IDC" (http://www.idc.com/getdoc.jsp?containerId=prUS23297412). ICD. 1 February 2012. .

[134] "The Wave of the Future" (http://www.underconsideration.com/brandnew/archives/the_wave_of_the_future.php). *Brand New: Opinions on Corporate and Brand Identity Work*. UnderConsideration LLC. 25 March 2007. . Retrieved 14 May 2008.

[135] "Reviews – 2007 – Nokia Siemens Networks" (http://www.identityworks.com/reviews/2007/Nokia_Siemens.htm). *Identityworks*. 2007. . Retrieved 14 May 2008.

[136] "Nokia Leadership Team" (http://www.nokia.com/A4126335). Nokia Corporation. April 2007. . Retrieved 14 May 2008.

[137] "Board of Directors" (http://www.nokia.com/A4126350). Nokia Corporation. April 2007. . Retrieved 14 May 2008.

[138] "Audit Committee Charter at Nokia" (http://www.nokia.com/NOKIA_COM_1/About_Nokia/Sidebars_new_concept/Board_charters/audit_charter.pdf) (PDF). Nokia Corporation. 2007. . Retrieved 14 May 2008.

[139] "Personnel Committee Charter at Nokia" (http://www.nokia.com/NOKIA_COM_1/About_Nokia/Sidebars_new_concept/Board_charters/personnel_charter_2007.pdf) (PDF). Nokia Corporation. 2007. . Retrieved 14 May 2008.

[140] "Corporate Governance and Nomination Committee Charter at Nokia" (http://www.nokia.com/NOKIA_COM_1/About_Nokia/Sidebars_new_concept/Board_charters/CG_Charter_2008_Final_20080123.pdf) (PDF). Nokia Corporation. 2008. . Retrieved 14 May 2008.

[141] "Committees of the Board" (http://www.nokia.com/link?cid=EDITORIAL_4207). Nokia Corporation. May 2007. . Retrieved 14 May 2008.

[142] Virkkunen, Johannes (29 September 2006). "New Finnish Companies Act designed to increase Finland's competitiveness" (http://www.lmr.fi/publications/companies_act_290906.pdf) (PDF). *LMR Attorneys Ltd. (Luostarinen Mettälä Räikkönen)*. . Retrieved 14 May 2008.

[143] "Articles of Association" (http://www.nokia.com/NOKIA_COM_1/About_Nokia/Company/Corporate_Governance/Articles_of_Association/Nokia_Articles_of_Association_10052007.pdf) (PDF). Nokia Corporation. 10 May 2007. . Retrieved 14 May 2008.

[144] "Corporate Governance Guidelines at Nokia" (http://www.nokia.com/NOKIA_COM_1/About_Nokia/Sidebars_new_concept/Board_charters/corporate_governance_guideline_sep06.pdf) (PDF). Nokia Corporation. 2006. . Retrieved 14 May 2008.

[145] "Suomalaisten yritysten ylin johto" (http://www.kolumbus.fi/taglarsson/dokumentit/yritys.htm) (in Finnish). . Retrieved 20 March 2009.

[146] "Nokia Research Center" (http://www.nokia.com/NOKIA_COM_1/Press/twwln/press_kit/Nokia_Research_Center_Press_Backgrounder_October_2007.pdf) (PDF). Nokia Corporation. October 2007. . Retrieved 14 May 2008.

[147] "About NRC – Nokia Research Center" (http://research.nokia.com/aboutus/index.html). Nokia Corporation. . Retrieved 17 March 2009.

[148] "NRC Locations – Nokia Research Center" (http://research.nokia.com/locations/index.html). Nokia Corporation. . Retrieved 17 March 2009.

[149] "INdT – Instituto Nokia de Tecnologia" (http://www.indt.org.br/). Nokia Corporation. . Retrieved 17 March 2009.

[150] "Production units" (http://www.nokia.com/A4149133). Nokia Corporation. June 2008. . Retrieved 14 May 2008.

[151] "Nokia despre sechestrul ANAF: Colaborăm pentru a ne asigura că situaţia e soluţionată satisfăcător" (http://www.mediafax.ro/economic/nokia-despre-sechestrul-anaf-colaboram-pentru-a-ne-asigura-ca-situatia-e-solutionata-satisfacator-8963720) (Press release). mediafax.ro. 12 November 2011. . Retrieved 12 November 2011.

[152] Kapanen, Ari (24 July 2007). "Ulkomaalaiset valtaavat pörssiyhtiöitä" (http://www.taloussanomat.fi/porssi-ja-raha/2007/07/24/Ulkomaalaiset+valtaavat+pörssiyhtiöitä/200717658/103) (in Finnish). *Taloussanomat*. . Retrieved 14 May 2008.

[153] "Nokia is no longer Finland's most valuable company" (http://www.phonearena.com/news/Nokia-is-no-longer-Finlands-most-valuable-company_id28750). phonearena.com. 4 April 2012. .

[154] Ali-Yrkkö, Jyrki (2001). "The role of Nokia in the Finnish Economy" (http://www.etla.fi/files/940_FES_01_1_nokia.pdf) (PDF). ETLA (The Research Institute of the Finnish Economy). . Retrieved 21 March 2009.

[155] Ali-Yrkkö, Jyrki (2010). "NOKIA AND FINLAND IN A SEA OF CHANGE" (http://www.etla.fi/files/2585_nokia_kirja_8_v2_kansineen.pdf). *ETLA – Research Institute of the Finnish Economy*. . Retrieved 12 November 2011.

[156] Guardian newspaper: Nokia cuts 4,000 jobs and moves smartphone manufacturing to Asia, 9 February 2012 (http://www.guardian.co.uk/technology/2012/feb/08/nokia-cuts-4000-manufacturing-jobs)

[157] "HS Archives" (http://www.hs.fi/arkisto/haku?pageNumber=1&order=FIFO&advancedSearch=&free=Connecting+People+Ove+Strandberg&date=year2003&depa=Kaikki+osastot&fromDay=0&fromMonth=0&fromYear=0&toDay=0&toMonth=0&toYear=0) (in Finnish). Helsingin Sanomat. 1 June 2003. . Retrieved 14 May 2008.

[158] "NOKIA | Connecting Pople 1992 Vector Logo (AI EPS)" (http://www.hdicon.com/vector-logos/nokia-connecting-pople-1992/). *HDicon.com*. . Retrieved 17 October 2010.

[159] Pitkänen, Juhani (Nokia's Art Director) (3 September 2007). "Nokia Strategic Marketing, Brand Identity" (http://www.nokia.com/A4126575). Nokia Corporation. . Retrieved 14 May 2008.

[160] "NOKIA | Connecting Pople new Vector Logo (AI EPS)" (http://www.hdicon.com/vector-logos/nokia-connecting-pople-new/). *HDicon.com*. . Retrieved 17 October 2010.

[161] "Erik Spiekermann – Furniture, Designs & Home Decor" (http://www.dwr.com/category/designers/r-t/erik-spiekermann.do). Design Within Reach. . Retrieved 7 January 2010.

[162] "Our New Typeface" (http://brandbook.nokia.com/blog/view/item62250/). Nokia Little Blog of Branding. . Retrieved 3 April 2012.

[163] http://thenextweb.com/eu/2012/04/04/poor-nokia-isnt-even-the-most-valuable-company-in-finland-anymore/
[164] http://media.corporate-ir.net/media_files/IROL/10/107224/Nokia_results2011Q2e.pdf
[165] http://www.theregister.co.uk/2012/04/19/nokia_earnings_ouch/
[166] http://thenextweb.com/insider/2012/05/14/nokia-is-getting-pummeled-stock-price-hits-staggering-low-after-a-6-nosedive/
[167] http://yle.fi/uutiset/nokia_share_price_in_nose_dive_to_2_euros/6094843
[168] "Nokia Way and values" (http://www.nokia.com/A4126303). Nokia Corporation. . Retrieved 14 May 2008.
[169] "dotMobi Investors" (http://web.archive.org/web/20080507214157/http://mtld.mobi/company/about/investors). dotMobi. Archived from the original (http://mtld.mobi/company/about/investors) on 7 May 2008. . Retrieved 14 May 2008.
[170] Haumont, Serge; Siren, Ritva. "dotMobi, a Key Enabler for the Mobile Internet" (http://research.nokia.com/files/Haumont-dotMobi.pdf) (PDF). *Nokia Research Center*. Nokia Corporation. . Retrieved 14 May 2008.
[171] "Nokia Ad Business" (http://www.adservice.nokia.com/faq.jsp#12). Nokia Corporation. . Retrieved 14 May 2008.
[172] Reardon, Marguerite (6 March 2007). "Nokia introduces mobile ad services" (http://news.com.com/2100-1039_3-6164800.html). *CNET News.com*. . Retrieved 14 May 2008.
[173] "Meet Ovi, the door to Nokia's Internet services" (http://www.nokia.com/A4136001?newsid=1149749) (Press release). Nokia Corporation. 29 August 2007. . Retrieved 7 April 2009.
[174] Niccolai, James (4 December 2007). "Nokia Lays Plan for More Internet Services" (http://www.nytimes.com/idg/IDG_002570DE00740E18002573A70046F2EF.html?ref=technology). *IDG News Service* (New York Times). . Retrieved 14 May 2008.
[175] "Ovi by Nokia" (http://www.nokia.com/NOKIA_COM_1/Press/Materials/White_Papers/pdf_files/backgrounders2008/Backgrounder_Ovi_by_Nokia.pdf) (PDF). Nokia Corporation. . Retrieved 7 April 2009.
[176] "Ovi Store opens for business" (http://www.nokia.com/press/press-releases/showpressrelease?newsid=1317441) (Press release). Nokia Corporation. 26 May 2009. . Retrieved 12 June 2009.
[177] Virki, Tarmo (18 March 2009). "Nokia to shutter its "Mosh" success story" (http://www.reuters.com/article/technologyNews/idUSTRE52H6AI20090318?pageNumber=1&virtualBrandChannel=0). *Reuters*. . Retrieved 14 July 2009.
[178] "Nokia launches new online magazine" (http://www.newstatesman.com/magazines/2010/03/online-magazine-nokia-ovi). NewStatesman. 23 March 2010. . Retrieved 29 March 2010.
[179] "Nokia – My Nokia" (http://europe.nokia.com/my-nokia). Nokia Corporation. . Retrieved 10 January 2012.
[180] "Nokia Retreats from Music Service" (http://www.yle.fi/uutiset/news/2011/01/nokia_retreats_from_music_service_2294706.html). YLE. 18 January 2011. . Retrieved 18 January 2011.
[181] Fields, Davis (17 December 2008). "Nokia Email service graduates as part of Nokia Messaging" (http://betalabs.nokia.com/blog/2008/12/17/nokia-email-service-graduates-as-part-of-nokia-messaging/). *Nokia Beta Labs*. Nokia Corporation. . Retrieved 16 March 2009.
[182] "Nokia Messaging: FAQ" (http://email.nokia.com/account/faq.action?change_locale=en). Nokia Corporation. . Retrieved 12 June 2009.
[183] Cellan-Jones, Rory (22 June 2009). "Hi-tech helps Iranian monitoring" (http://news.bbc.co.uk/1/hi/technology/8112550.stm). *BBC News*. . Retrieved 14 July 2009.
[184] Rhoads, Christopher; Chao, Loretta (22 June 2009). "Iran's Web Spying Aided By Western Technology" (http://online.wsj.com/article/SB124562668777335653.html#mod). *The Wall Street Journal* (Dow Jones & Company, Inc.): pp. A1. . Retrieved 14 July 2009.
[185] "Provision of Lawful Intercept capability in Iran" (http://www.nokiasiemensnetworks.com/global/Press/Press+releases/news-archive/Provision+of+Lawful+Intercept+capability+in+Iran.htm) (Press release). Nokia Siemens Networks. 22 June 2009. . Retrieved 14 July 2009.
[186] Kamali Dehghan, Saeed (14 July 2009). "Iranian consumers boycott Nokia for 'collaboration'" (http://www.guardian.co.uk/world/2009/jul/14/nokia-boycott-iran-election-protests). *The Guardian* (London: Guardian News and Media Limited). . Retrieved 27 July 2009.
[187] Ozimek, John (6 March 2009). "'Lex Nokia' company snoop law passes in Finland" (http://www.theregister.co.uk/2009/03/06/finland_nokia_snooping/). *The Register*. . Retrieved 27 July 2009.
[188] "Nokia Denies Threat to Leave Finland" (http://www.cellular-news.com/story/35783.php). *cellular-news*. 1 February 2009. . Retrieved 27 July 2009.
[189] Virki, Tarmo (18 January 2010). "SCENARIOS-What lies ahead in Nokia vs Apple legal battle" (http://www.reuters.com/article/idUSLDE60H05R20100118?type=marketsNews). *Reuters*. . Retrieved 25 January 2010.
[190] "The war of the Smartphones: Nokia's new patent suit against Apple" (http://pda-phone-reviews.in/latest-news/nokias-new-patent-suit-against-apple/). *Snartphone Reviews*. 6 January 2010. . Retrieved 25 January 2010.
[191] "Nokia's Patent Settlement With Apple Won't Help Much" (http://www.informationweek.com/news/personal-tech/smart-phones/230600172). 14 Jun 2011. . Retrieved 29 June 2011.
[192] Smith, Catharine (14 Jun 2011). "Apple Settles With Nokia In Patent Lawsuit" (http://www.huffingtonpost.com/2011/06/14/apple-nokia-patent-lawsuit-settlement_n_876499.html). *Huffington Post*. . Retrieved 29 June 2011.
[193] ben-Aaron, Diana; Pohjanpalo, Kati (14 Jun 2011). "Nokia Wins Apple Patent-License Deal Cash, Settles Lawsuits" (http://www.bloomberg.com/news/2011-06-14/nokia-apple-payments-to-nokia-settle-all-litigation.html). *Bloomberg*. . Retrieved 29 June 2011.
[194] "Guide to Greener Electronics – Greenpeace International" (http://www.greenpeace.org/international/en/campaigns/climate-change/cool-it/Guide-to-Greener-Electronics/). Greenpeace International. . Retrieved 14 November 2011.
[195] "Nokia – Where and how to recycle – Recycling – Environment" (http://www.nokia.com/environment/recycling/where-and-how-to-recycle). Nokia. . Retrieved 12 August 2010.

[196] "Global consumer survey reveals that majority of old mobile phones are lying in drawers at home and not being recycled" (http://www.nokia.com/press/press-releases/showpressrelease?newsid=1234291) (Press release). Nokia Corporation. 8 July 2008. . Retrieved 27 July 2009.

[197] "Nokia – Energy efficiency – Devices and services – Environment" (http://www.nokia.com/environment/we-energise/nokia-and-energy-efficiency). Nokia. . Retrieved 12 August 2010.

[198] "Nokia – Energy saving targets – Environmental strategy – Strategy and reports – Environment" (http://www.nokia.com/environment/strategy-and-reports/environmental-strategy/energy-saving-targets). Nokia. . Retrieved 12 August 2010.

[199] "Nokia" (http://www.greenpeace.org/international/en/campaigns/toxics/electronics/Guide-to-Greener-Electronics/companies/Nokia/). Greenpeace International. . Retrieved 12 August 2010.

[200] "Materials and substances" (http://www.nokia.com/environment/we-create/materials-and-substances). Nokia Corporation. . Retrieved 12 August 2010.

[201] "Eco declarations" (http://www.nokia.com/environment/we-create/devices-and-accessories/eco-declarations). Nokia Corporation. . Retrieved 27 July 2009.

[202] Rubio, Jenalyn (12 April 2008). "Tech Goes Greener" (http://www.pcworld.com/article/id,144482-c,recycling/article.html). *Computerworld Philippines* (PC World). . Retrieved 14 May 2008.

[203] "Nokia Remade Concept Phone goes Green" (http://www.mobiletor.com/2008/04/09/nokia-remade-concept-phone-goes-green/). Mobiletor. 9 April 2008. . Retrieved 14 May 2008.

[204] "Open Innovation – Nokia Research Center" (http://research.nokia.com/openinnovation). Nokia Corporation. . Retrieved 1 April 2009.

[205] "Nokia is India's most trusted brand : Brand Trust" (http://profit.ndtv.com/news/show/nokia-is-india-s-most-trusted-brand-brand-trust-136783). profit.ndtv.com. 19 January 2011. . Retrieved 21 November 2011.

Further reading

Title	Author	Publisher	Year	Length	ISBN
Winning Across Global Markets: How Nokia Creates Strategic Advantage in a Fast-Changing World	Dan Steinbock	Jossey-Bass / Wiley	May 2010	304 pp	ISBN 978-0-470-33966-4
Nokia: The Inside Story	Martti Häikiö	FT / Prentice Hall	October 2002	256 pp	ISBN 0-273-65983-9
Work Goes Mobile: Nokia's Lessons from the Leading Edge	Michael Lattanzi, Antti Korhonen, Vishy Gopalakrishnan	John Wiley & Sons	January 2006	212 pp	ISBN 0-470-02752-5
Mobile Usability: How Nokia Changed the Face of the Mobile Phone	Christian Lindholm, Turkka Keinonen, Harri Kiljander	McGraw-Hill Companies	June 2003	301 pp	ISBN 0-07-138514-2
Business The Nokia Way: Secrets of the World's Fastest Moving Company	Trevor Merriden	John Wiley & Sons	February 2001	168 pp	ISBN 1-84112-104-5
The Nokia Revolution: The Story of an Extraordinary Company That Transformed an Industry	Dan Steinbock	AMACOM Books	April 2001	375 pp	ISBN 0-8144-0636-X

External links

- Official Nokia international website (http://www.nokia.com/)

QWERTY

A QWERTY keyboard on a laptop computer

QWERTY (/ˈkwɜrti/) is the most common modern-day keyboard layout. The name comes from the first six letters (keys) appearing in the top left letter row of the keyboard, read left to right: Q-W-E-R-T-Y. The QWERTY design is based on a layout created for the Sholes and Glidden typewriter and sold to Remington in the same year, when it first appeared in typewriters. It became popular with the success of the Remington No. 2 of 1878, and remains in use on electronic keyboards due to the network effect of a standard layout and a belief that alternatives fail to provide very significant advantages.[1] The use and adoption of the QWERTY keyboard is often viewed as one of the most important case studies in open standards because of the widespread, collective adoption and use of the product, particularly in the United States.[2]

History and purposes

Keys are arranged on diagonal columns, to give space for the levers.

This layout was devised and created in the early 1870s by Christopher Latham Sholes, a newspaper editor and printer who lived in Milwaukee. With the assistance of his friends Carlos Glidden and Samuel W. Soule he built an early writing machine for which a patent application was filed in October 1867.[3]

The first model constructed by Sholes used a piano-like keyboard with two rows of characters arranged alphabetically as follows:[3]

```
- 3 5 7 9 N O P Q R S T U V W X Y Z
 2 4 6 8 . A B C D E F G H I J K L M
```

His "Type Writer" had two features which made jams a serious issue. Firstly, characters were mounted on metal arms or typebars, which would clash and jam if neighboring arms were depressed at the same time or in rapid succession.[4] Secondly, its printing point was located beneath the paper carriage, invisible to the operator, a so-called "up-stroke" design. Consequently, jams were especially serious, because the typist could only discover the mishap by raising the carriage to inspect what he had typed. The solution was to place commonly used letter-pairs (like "th" or "st") so that their typebars were not neighboring, avoiding jams. A popular myth is that QWERTY was designed to "slow down" typists though this is incorrect – it was designed to prevent jams[4] *while* typing at speed, allowing typists to type *faster*.[5]

Sholes struggled for the next five years to perfect his invention, making many trial-and-error rearrangements of the original machine's alphabetical key arrangement. His study of letter-pair frequency by educator Amos Densmore, brother of the financial backer James Densmore, is believed to have influenced the arrangement of letters, but called in question.[6]

In November 1868 he changed the arrangement of the latter half of the alphabet, O to Z, right-to-left.[7] In April 1870 he arrived at a four-row, upper case keyboard approaching the modern QWERTY standard, moving six vowels,

A, E, I, O, U, and Y, to the upper row as follows:[8]

2 3 4 5 6 7 8 9 -
A E I . ? Y U O ,
B C D F G H J K L M
Z X W V T S R Q P N

In 1873 Sholes's backer, James Densmore, succeeded in selling manufacturing rights for the Sholes & Glidden Type-Writer to E. Remington and Sons, and within a few months the keyboard layout was finalized by Remington's mechanics. The keyboard ultimately presented to Remington was arranged as follows:[9]

2 3 4 5 6 7 8 9 - ,
Q W E . T Y I U O P
Z S D F G H J K L M
A X & C V B N ? ; R

After it purchased the device, Remington made several adjustments which created a keyboard with what is essentially the modern QWERTY layout. Their adjustments included placing the "R" key in the place previously allotted to the period key (this has been claimed to be done with the purpose of enabling salesmen to impress customers by pecking out the brand name "TYPE WRITER" from one keyboard row but this claim is unsubstantiated[9]). Vestiges of the original alphabetical layout remained in the "home row" sequence DFGHJKL.[10]

The QWERTY layout became popular with the success of the Remington No. 2 of 1878, the first typewriter to include both upper and lower case letters, via a shift key.

Much less commented-on than the order of the keys is that the keys are not on a grid, but rather that each column slants diagonally; this is because of the mechanical linkages – each key being attached to a lever, and hence the offset prevents the levers from running into each other – and has been retained in most electronic keyboards. Some keyboards, such as the Kinesis, retain the QWERTY layout but arrange the keys in vertical columns, to reduce unnecessary lateral finger motion.[11]

Differences from modern layout

Substituting characters

The QWERTY layout depicted in Sholes's 1878 patent includes a few differences from the modern layout, most notably in the absence of the numerals 0 and 1, with each of the remaining numerals shifted one position to the left of their modern counterparts. The letter M is located at the end of the third row to the right of the letter L rather than on the fourth row to the right of the N, the letters X and C are reversed, and most punctuation marks are in different positions or are missing entirely.[12] 0 and 1 were omitted to simplify the design and reduce the manufacturing and maintenance costs; they were chosen specifically because they were "redundant" and could be recreated using other keys. Typists who learned on these machines learned the habit of using the uppercase letter I (or lowercase letter L) for the digit one, and the uppercase O for the zero.[13]

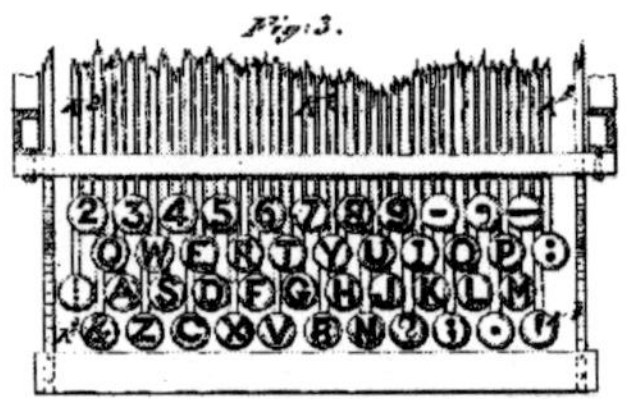
Latham Sholes's 1878 QWERTY keyboard layout

Combined characters

In early designs, some characters were produced by printing two symbols with the carriage in the same position. For instance, the exclamation point, which shares a key with the numeral 1 on modern keyboards, could be reproduced by using a three-stroke combination of an apostrophe, a backspace, and a period. A semicolon (;) was produced by printing a comma (,) over a colon (:). As the backspace key is slow in simple mechanical typewriters (the carriage was heavy and optimized to move in the opposite direction), a more professional approach was to block the carriage by pressing and holding the space bar while printing all characters that needed to be in a shared position. To make this possible, the carriage was designed to advance forward only after releasing the space bar.

The 0 key was added and standardized in its modern position early in the history of the typewriter, but the 1 and exclamation point were left off some typewriter keyboards into the 1970s.[14]

Contemporary alternatives

There was no particular technological requirement for the QWERTY layout,[9] since at the time there were ways to make a typewriter without the "up-stroke" typebar mechanism that had required it to be devised. Not only were there rival machines with "down-stroke" and "frontstroke" positions that gave a visible printing point, the problem of typebar clashes could be circumvented completely: examples include Thomas Edison's 1872 electric print-wheel device which later became the basis for Teletype machines; Lucien Stephen Crandall's typewriter (the second to come onto the American market) whose type was arranged on a cylindrical sleeve; the Hammond typewriter of 1887 which used a semi-circular "type-shuttle" of hardened rubber (later light metal); and the Blickensderfer typewriter of 1893 which used a type wheel. The early Blickensderfer's "Ideal" keyboard was also non-QWERTY, instead having the sequence "DHIATENSOR" in the home row, these 10 letters being capable of composing 70% of the words in the English language.[15]

Properties

Alternating hands while typing is a desirable trait in a keyboard design, since while one hand is typing a letter, the other hand can get in position to type the next letter. Thus, a typist may fall into a steady rhythm and type quickly. However, when a string of letters is done with the same hand, the chances of stuttering are increased and a rhythm can be broken, thus decreasing speed and increasing errors and fatigue. In the QWERTY layout many more words can be spelled using only the left hand than the right hand. In fact, thousands of English words can be spelled using only the left hand, while only a couple of hundred words can be typed using only the right hand. In addition, most typing strokes are done with the left hand in the QWERTY layout. This is helpful for left-handed people but to the disadvantage of right-handed people.[16]

Effects

People tend to give slightly more negative connotations to words that are typed with the left hand on the QWERTY keyboard, including its variants in several languages.[17]

Computer keyboards

The first computer terminals such as the Teletype were typewriters that could produce and be controlled by various computer codes. These used the QWERTY layouts, and added keys such as escape (ESC) which had special meanings to computers. Later keyboards added function keys and arrow keys. Since the standardization of PC-compatible computers and Windows after the 1980s, most full-sized computer keyboards have followed this standard (see drawing at right). This layout has a separate numeric keypad for data entry at the right, 12 function keys across the top, and a cursor section to the right and center with keys for Insert, Delete, Home, End, Page Up, and Page Down with cursor arrows in an inverted-T shape.

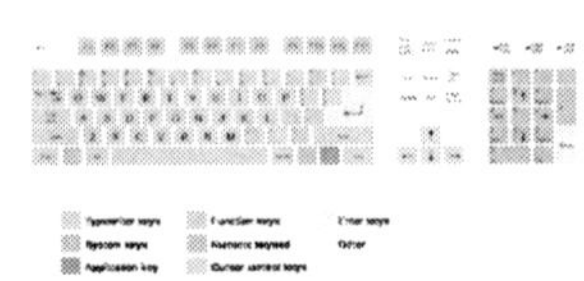

The standard QWERTY keyboard layout used in the US. Some countries, such as the UK and Canada, use a slightly different QWERTY (the @ and " are switched in the UK); see keyboard layout.

Diacritical marks and international variants

Different computer operating systems have methods of support for input of different languages such as Chinese, Hebrew or Arabic. QWERTY is designed for English, a language without any diacritical marks. QWERTY keyboards meet issues when having to type an accent. Until recently, no norm was defined for a standard QWERTY keyboard layout allowing the typing of accented characters, apart from the US-International layout.

Depending on the operating system and sometimes the application program being used, there are many ways to generate Latin characters with accents.

UK-Extended Layout

Microsoft Windows XP SP2 and above provide the UK-Extended layout that behaves exactly the same as the standard UK layout for all the characters it can generate, but can additionally generate a number of diacritical marks, useful when working with text in other languages (including Welsh - a UK language). Not all combinations work on all keyboards.

- acute accents (e.g. á) on a,e,i,o,u,w,y,A,E,I,O,U,W,Y are generated by pressing the AltGr key together with the letter, or AltGr and apostrophe, followed by the letter (see note below);
- grave accents (e.g. è) on a,e,i,o,u,w,y,A,E,I,O,U,W,Y are generated by pressing the backquote (`) [which is now a dead key], then the letter;
- circumflex (e.g. â) on a,e,i,o,u,w,y,A,E,I,O,U,W,Y is generated by AltGr and 6, followed by the letter;
- diaeresis or umlaut (e.g. ö) on a,e,i,o,u,w,y,A,E,I,O,U,W,Y is generated by AltGr and 2, then the letter;
- tilde (e.g. ã) on a,n,o,A,N,O is generated by AltGr and #, then the letter;
- cedilla (e.g. ç) under c,C is generated by AltGr and the letter.

These combinations are designed to be easy to remember, as the circumflex accent (e.g. â) is similar to a caret (^), printed above the 6 key; the diaeresis (e.g. ö) is similar to the double-quote (") above 2 on the UK keyboard; the tilde (~) is printed on the same key as the #.

Like US-International, UK-Extended does not cater for many languages written with Latin characters, including Romanian and Turkish, or any using different character sets such as Greek and Russian.

Notes:

- The AltGr and letter method used for acutes and cedillas does not work for applications which assign shortcut menu functions to these key combinations. For acute accents the AltGr and apostrophe method should be used.

International variants

Minor changes to the arrangement are made for other languages.

Alternatives to QWERTY

Several alternatives to QWERTY have been developed over the years, claimed by their designers and users to be more efficient, intuitive and ergonomic. Nevertheless, none has seen widespread adoption, due partly to the sheer dominance of available keyboards and training.[18] Although studies have shown the superiority in typing speed afforded by alternative keyboard layouts[19] economists Stan Liebowitz and Stephen E Margolis have claimed that these studies are flawed and more rigorous studies are inconclusive as to whether they actually offer any real benefits.[1] The most widely used such alternative is the Dvorak Simplified Keyboard; another increasingly popular alternative is Colemak, which is based partly on QWERTY and is therefore easier for an existing QWERTY typist to learn while offering several optimisations.[20] Most modern computer operating systems support this and other alternative mappings with appropriate special mode settings, but few keyboards are manufactured with keys labeled according to this standard.

Half QWERTY

A half QWERTY keyboard is a combination of an alpha-numeric keypad and a QWERTY keypad, designed for mobile phones.[21] In a half QWERTY keyboard, two characters share the same key, which reduces the number of keys and increases the surface area of each key, useful for mobile phones that have little space for keys.[21] It means that 'Q' and 'W' will share the same key and the user has to press the key once to type 'Q' and twice to type 'W'.

The Nokia E55 uses a half QWERTY keyboard layout.

Displaced QWERTY

Also designed for mobile devices, the displaced QWERTY layout allows for the increase of button area by over 40% while keeping the same candybar form factor. Entering, spacing and deleting are handled by gestures over the text area, reducing the keyboard's screen footprint. The layout is essentially a rearrangement of keys on the right half of the keyboard under those on the left and, as such, should present a gentler learning curve to touch typists. It was first seen on the iPhone application "LittlePad".

References

[1] Liebowitz, Stan; Margolis, Stephen E. (1990), "The Fable of the Keys", *Journal of Law and Economics* **33** (1): 1–26, doi:10.1086/467198

[2] "Casson and Ryan, Open Standards, Open Source Adoption in the Public Sector, and Their Relationship to Microsoft's Market Dominance" (http://papers.ssrn.com/sol3/papers.cfm?abstract_id=1656616). Papers.ssrn.com. . Retrieved 2011-01-31.

[3] US 79868 (http://worldwide.espacenet.com/textdoc?DB=EPODOC&IDX=US79868), Sholes, C. Latham; Carlos Glidden & Samuel W. Soule, "Improvement in Type-writing Machines", issued July 14, 1868

[4] Rehr, Darryl, *Why QWERTY was Invented* (http://home.earthlink.net/~dcrehr/whyqwert.html),

[5] Rehr, Darryl. "Consider QWERTY" (http://home.earthlink.net/~dcrehr/whyqwert.html). . Retrieved 12 December 2011. "QWERTY's effect, by reducing those annoying clashes, was to speed up typing rather than slow it down."

[6] Koichi Yasuoka: The Truth of QWERTY (http://yasuoka.blogspot.com/2006/08/sholes-discovered-that-many-english.html), entry dated August 01, 2006.

[7] Koichi and Motoko Yasuoka: Myth of QWERTY Keyboard, Tokyo: NTT Publishing, 2008. pp.12-20 (http://books.google.com/books?id=tEsAMggMKoMC&pg=PA8)

[8] Koichi and Motoko Yasuoka: Myth of QWERTY Keyboard, Tokyo: NTT Publishing, 2008. pp.24-25 (http://books.google.com/books?id=tEsAMggMKoMC&pg=PA20)

[9] Koichi and Motoko Yasuoka: On the Prehistory of QWERTY (http://kanji.zinbun.kyoto-u.ac.jp/~yasuoka/publications/PreQWERTY.html), ZINBUN, No.42, pp.161-174, 2011.

[10] David, Paul A. (1985), "Clio and the Economics of QWERTY", *American Economic Review* (American Economic Association) **75** (2): 332–337, JSTOR 1805621

[11] Kinesis – Ergonomic Benefits of the Contoured Keyboard (http://www.kinesis-ergo.com/benefits.htm) – Vertical key layout

[12] US 207559 (http://worldwide.espacenet.com/textdoc?DB=EPODOC&IDX=US207559), Sholes, Christopher Latham, issued August 27, 1878

[13] Weller, Charles Edward (1918), *The early history of the typewriter* (http://www.archive.org/details/earlyhistorytyp00wellgoog), La Porte, Indiana: Chase & Shepard, printers,

[14] See for example the Olivetti Lettera 36 (http://www.mrmartinweb.com/type.htm#olivetti), introduced in 1972

[15] Shermer, Michael (2008). *The mind of the market*. Macmillan. p. 50. ISBN 0-8050-7832-0.

[16] Diamond, Jared (April 1997), "The Curse of QWERTY" (http://discovermagazine.com/1997/apr/thecurseofqwerty1099/), *Discover*, , retrieved 2009-04-29, "More than 3,000 English words utilize QWERTY's left hand alone, and about 300 the right hand alone."

[17] "Keyboards and vocabulary: Hunting for the right words" (http://www.economist.com/blogs/johnson/2012/03/keyboards-and-vocabulary?fsrc=scn/fb/wl/bl/huntingfortherightwords). *The Economist*. . Retrieved 14 March 2012.

[18] Gould, Stephen Jay (1987) "The Panda's Thumb of Technology." (http://books.google.com/books?id=pzj90slTTEIC&pg=PA59) *Natural History* 96 (1): 14-23; Reprinted in *Bully for Brontosaurus*. New York: W.W. Norton. 1992, pp. 59-75.

[19] Paul David, "Understanding the economics of QWERTY: the necessity of history", *Economic history and the modern economist*, 1986

[20] Krzywinski, Martin. "Colemak - Popular Alternative" (http://mkweb.bcgsc.ca/carpalx/?colemak). *Carpalx - keyboard layout optimizer*. Canada's Michael Smith Genome Sciences Centre. . Retrieved 2010-02-04.

[21] "Half-QWERTY keyboard layout - Mobile terms glossary" (http://www.gsmarena.com/glossary.php3?term=half-qwerty-keyboard). GSMArena.com. . Retrieved 2011-01-31.

External links

- Article on QWERTY and Path Dependence from EH.NET's Encyclopedia (http://eh.net/encyclopedia/article/puffert.path.dependence)
- QWERTY Keyboard History (http://www.ideafinder.com/history/inventions/qwerty.htm)
- QWERTY Keyboard in Mobiles (http://www.bakwaash.com/2011/07/05/mobile-phone-termonologies/)

Entry_Level

For the general term, see entry-level at Wiktionary.

Entry Level is the lowest level in the National Qualifications Framework in England, Wales, and Northern Ireland. Qualifications at this level recognise basic knowledge and skills and the ability to apply learning in everyday situations under direct guidance or supervision. Learning at this level involves building basic knowledge and skills and is not usually geared towards specific occupations.

Entry Level qualifications can be taken at three levels (Entry 1, Entry 2 and Entry 3[1]) and are available on a broad range of subjects. They are targeted at a range of learners, including adult learners, candidates on taster sessions, underachievers and ones with learning difficulties.[2]

The level after Entry Level in the National Qualifications Framework is Level 1, which includes GCSE grades D-G and Level 1 DiDA.

Examples of Entry Level qualifications

- Entry Level Certificate
- Entry Level Functional Skills

References

[1] "Edexcel : Qualifications : Entry Level Certificate" (http://www.edexcel.org.uk/quals/elc/). Edexcel. . Retrieved 2008-05-29.
[2] "OCR > Qualifications > Entry Level" (http://www.ocr.org.uk/qualifications/entrylevel/). OCR. . Retrieved 2008-05-29

Secure_Digital

SD, SDHC, SDXC

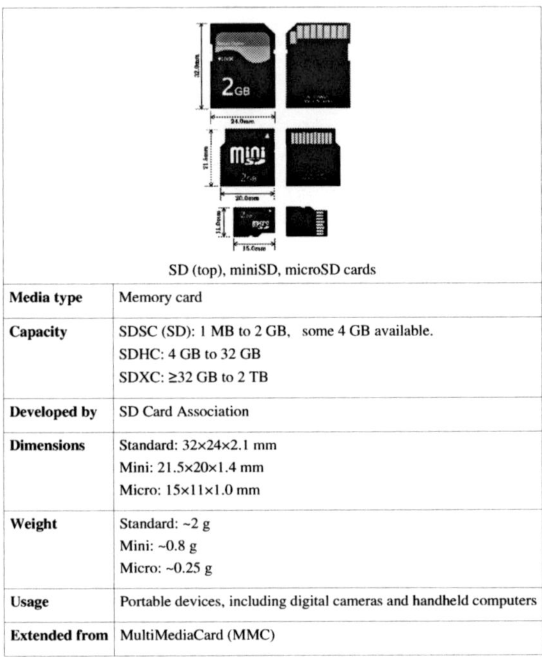

SD (top), miniSD, microSD cards

Media type	Memory card
Capacity	SDSC (SD): 1 MB to 2 GB, some 4 GB available. SDHC: 4 GB to 32 GB SDXC: ≥32 GB to 2 TB
Developed by	SD Card Association
Dimensions	Standard: 32×24×2.1 mm Mini: 21.5×20×1.4 mm Micro: 15×11×1.0 mm
Weight	Standard: ~2 g Mini: ~0.8 g Micro: ~0.25 g
Usage	Portable devices, including digital cameras and handheld computers
Extended from	MultiMediaCard (MMC)

Secure Digital or (**SD**) is a non-volatile memory card format for use in portable devices. The Secure Digital standard is maintained by the SD Card Association (SDA). SD technologies have been implemented in more than 400 brands across dozens of product categories and more than 8,000 models.[1]

The Secure Digital format includes four card families available in three different form factors. The four families are the original, Standard-Capacity (SDSC), the High-Capacity (SDHC), the eXtended-Capacity (SDXC), and the SDIO, which combines input/output functions with data storage.[2][3][4] The three form factors are the original size, the "mini" size, and the "micro" size (see illustration). There are many combinations of form factors and device families.

Electrically passive adapters allow the use of a smaller card in a host device built to hold a larger card.

Host devices that comply with newer versions of the specification provide backward compatibility and accept older SD cards, but older host devices do not recognize newer cards. The SDA uses several trademarked logos to enforce compliance with its specifications and assure users of compatibility.[5] This article explains several factors that can prevent the use of a newer SD card:

- A newer card may offer greater capacity than the host device can handle.
- A newer card may use a file system the host device cannot navigate.
- Use of an SDIO card requires the host device be designed for the input/output functions the card provides.
- The organization of the card was changed starting with the SDHC family.
- Some vendors produced SDSC cards above 1 GB before the SDA had standardized a method of doing so.

Types of cards

The SDA extended the SD specification in various ways:

- It defined electrically identical cards in smaller sizes: miniSD and microSD (originally named TransFlash or TF). Smaller cards are usable in larger slots through use of a passive adapter. By comparison, Reduced Size MultiMediaCards (RS-MMCs) are simply shorter MMCs and can be used in MMC slots by use of a physical extender.
- It defined higher-capacity cards, some with faster speeds and added capabilities: SDHC (Secure Digital High Capacity) and SDXC (Secure Digital eXtended Capacity). These cards redefine the interface so that they cannot be used in older host devices.
- It defined an SDIO card family that provides input-output functions and may also provide memory functions. These cards are only fully functional in host devices designed to support their input-output functions.

Physical size

The SD card specification defines three physical sizes. The SD and SDHC families are available in all three sizes, but the SDXC family is not available in the mini size, and the SDIO family is not available in the micro size.

Size comparison of families: SD, miniSD, microSD

Standard size

- SD (SDSC), SDHC, SDXC, SDIO
- 32 mm × 24 mm × 2.1 mm
- 32 mm × 24 mm × 1.4 mm (as thin as MMC) for **Thin SD** (rare)

Mini size

- miniSD, miniSDHC, miniSDIO
- 21.5 mm × 20 mm × 1.4 mm

Micro size

The microSD form factor is the smallest memory card format currently available.[6]

- microSD, microSDHC, microSDXC
- 15 mm × 11 mm × 1.0 mm

SDHC

The Secure Digital High Capacity (SDHC) format, defined in Version 2.0 of the SD specification, supports cards with capacities up to 32 GB.[1] The SDHC trademark is licensed to ensure compatibility.[7]

16GB SDHC Class 6

SDHC cards are physically and electrically identical to standard-capacity SD cards (SDSC). The major compatibility issues between SDHC and SDSC cards are the redefinition of the Card-Specific Data (CSD) register in Version 2.0 (see below), and the fact that SDHC cards are shipped preformatted with the FAT32 file system.

Host devices that accept SDHC cards are required to accept SDSC cards.[1] However, host devices designed for SDSC do not recognize SDHC or SDXC memory cards, although some devices can do so through a firmware upgrade.[8] Older operating systems require patches to support SDHC. For instance, Microsoft Windows XP before SP3 requires a patch to support access to SDHC cards.[9] Windows Vista SP1 also requires a later service pack.[10][11]

SDXC

The Secure Digital eXtended Capacity (SDXC) format supports cards up to 2 TB (2048 GB), compared to a limit of 32 GB for SDHC cards in the SD 2.0 specification.

Official SDXC logo.

History

SDXC was announced at Consumer Electronics Show (CES) 2009 (January 7–10, 2009). At the same show, SanDisk and Sony also announced a comparable Memory Stick XC variant with the same 2 TB maximum as SDXC,[12] and Panasonic announced plans to produce 64 GB SDXC cards.[13]

On March 6, 2009, Pretec introduced the first SDXC card,[14] a 32 GB card with a read/write speed of 400 Mbit/s. But it was not until early 2010 that compatible host devices came onto the market, including Sony's Handycam HDR-CX55V camcorder, Canon's EOS Rebel T2i Digital SLR camera,[15] a USB card reader from Panasonic, and an integrated SDXC card reader from JMicron.[16] The earliest laptops to integrate SDXC card readers relied on a USB 2.0 bus, which does not have the bandwidth to support SDXC at full speed.[17]

Also in early 2010, commercial SDXC cards appeared from Toshiba (64 GB),[18][19] Panasonic (64 GB and 48 GB),[20] and SanDisk (64 GB).[21] In early 2011, Centon Electronics, Inc. (64 GB and 128 GB) and Lexar (128 GB)[22] began shipping SDXC cards rated at Speed Class 10. Pretec offered cards from 8 GB to 128 GB rated at Speed Class 16.[23]

In September 2011, SanDisk released a 64 GB microSDXC card.[24] Kingmax released a comparable product in 2011.[25]

Compatibility with SDHC

SDXC host devices accept all previous families of SD memory cards.[26] Conversely, SDHC host devices will accept SDXC cards that follow Version 3.0, since the interface is identical,[3] but the following issues may affect usability:

- SDXC cards are pre-formatted with Microsoft's proprietary and patented exFAT file system, which the host device might not support. Since Microsoft does not publish the specifications of exFAT and its use requires a non-free license, many alternative or older operating systems do not support exFAT for technical or legal reasons. The use of exFAT on some SDXC cards may render SDXC unsuitable as a universal exchange medium, as an

SDXC card that uses exFAT would not be usable in all host devices. Since the FAT32 file system supports volumes up to the SDXC's maximum theoretical capacity of 2 TB as well, a user could reformat an SDXC card to use FAT32 for greater portability between computers (see below). FAT32-formatted SDXC cards can be used in a host device built for SDHC if the host device can handle 64GB and larger volumes.

- SDHC host devices will not test the new capability bits defined for SDXC 4.0 cards. It will therefore not be able to use the new features of SDXC, such as transfer speeds above UHS104 (104MB/s).

Host operating system support

Microsoft Windows versions that support SDXC are: Windows 7, Windows Vista SP1+,[3] Windows XP SP2 or SP3 with KB955704,[27] Windows Server 2008 SP1+, Windows Server 2003 SP2 or SP3 with KB955704, and Windows CE 6.0 and higher.

Apple Mac OS X versions that support SDXC cards and exFAT are Mac OS X Snow Leopard 10.6.5 or later including OS X Lion 10.7.[28][29]

BSD and Linux systems that support SDHC cards also support SDXC cards that contain a compatible file system, but they normally do not support the proprietary exFAT file system that is installed on SDXC cards, because of patent issues. However, Google Code contains an implementation.[30] The user may reformat the card to contain a different file system (see below).

SDIO

A SDIO (Secure Digital Input Output) card is an extension of the SD specification to cover I/O functions. Host devices that support SDIO (typically PDAs like the Palm Treo, but occasionally laptops or mobile phones) can use the SD slot to support GPS receivers, modems, barcode readers, FM radio tuners, TV tuners, RFID readers, digital cameras, and interfaces to Wi-Fi, Bluetooth, Ethernet, and IrDA. Many other SDIO devices have been proposed, but it is now more common for I/O devices to connect using the USB interface.

Camera using the SDIO interface to connect to some HP iPAQ devices

SDIO cards support most of the memory commands of SD cards. SDIO cards can be structured as 8 logical cards, although currently, the typical way that an SDIO card uses this capability is to structure itself as one I/O card and one memory card.

Host support for SDIO

The SDIO and SD interfaces are mechanically and electrically identical. Host devices built for SDIO cards generally accept SD memory cards without I/O functions. However, the reverse is not true, because host devices need suitable drivers and applications to support the card's I/O functions. For example, an HP SDIO camera usually does not work with PDAs that do not list it as an accessory. Inserting an SDIO card into any SD slot causes no physical damage nor disruption to the host device, but users may be frustrated that the SDIO card does not function fully when inserted into a seemingly compatible slot. (Bluetooth devices exhibit comparable compatibility issues, although to a lesser extent thanks to standardized Bluetooth profiles.)

Features

Card security (non-DRM)

Read-only cards

The host device can command the SD card to become read-only (to reject subsequent commands to write information to it). There are several different ways to achieve this, one of which is irreversible.

Using a read-only device in Microsoft Windows may produce errors, as even when merely reading a file, Windows tries to write a timestamp to the device.

Write protection tab

When looking at the card from the top, the right side (the side with the beveled corner) must be notched.

On the left side, there may be a write-protection notch. If the notch is omitted, the card can be read and written. If the card is notched, it is read-only, except that the notch may be partly covered by a sliding tab. In this case, the user can slide the tab upward (toward the contacts) to declare the card read/write, or downward to declare it read-only.

The presence of a notch, and the presence and position of a tab, have no effect on the SD card's operation. A host device that supports write protection should refuse to write to an SD card that is designated read-only in this way. Some host devices do not support write protection, which is an optional feature of the SD specification. Host devices that do obey a read-only indication may give the user a way to override it. The miniSD and microSD formats do not support write protection with the notch-and-tab method.

Typical cards sold as a medium for protected content are permanently marked read-only by having a notch and no sliding tab.

Card password

A host device can lock an SD card using a password of up to 16 bytes, typically supplied by the user. A locked card interacts normally with the host device except that it rejects commands to read and write data. A locked card can be unlocked only by providing the same password. The host device can, after supplying the old password, specify a new password or disable locking. Without the password (typically, in the case that the user forgets the password), the host device can command the card to erase all the data on the card for future re-use (except card data under DRM), but there is no way to gain access to the existing data.

DRM features

All SD cards incorporate a digital rights management (DRM) scheme. Roughly 10% of the storage capacity of an SD card is not available to the user, but is used by the on-card processor to verify the identity of an application program that it will then allow to read protected content. The card prohibits other accesses, such as users trying to make copies of protected files.

The DRM scheme embedded in the SD cards is the Content Protection for Recordable Media (CPRM or CPPM) specification of the 4C Entity, which features the Cryptomeria cipher (also termed *C2*). The specification is kept secret and is accessible only to licensees. The scheme has not been broken or hacked, but this feature of SD cards is rarely used to protect content. DVD-Audio uses the same DRM scheme.

Windows Media files can be DRM-encoded so as to make use of the SD card's DRM abilities.

Windows Phone 7 devices use SD cards designed to be accessed only by the phone manufacturer or mobile provider. An SD card inserted into the phone underneath the battery compartment becomes locked "to the phone with an automatically generated key" so that "the SD card cannot be read by another phone, device, or PC".[31] Symbian devices, however, are some of the very few which can perform the necessary low-level format operations on locked SD cards. It is therefore possible to use a device such as the Nokia N8 to reformat the card for subsequent use in other devices.[32]

A *Super Digital* product formerly[33] made by Super Talent was essentially the SD card without the DRM features.

Vendor enhancements

Vendors have sought to differentiate their products in the market through various vendor-specific features:

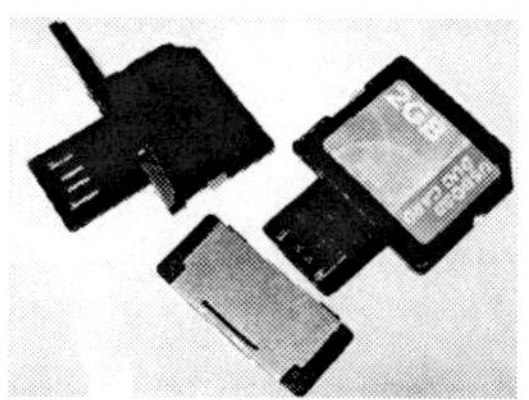

SD cards with dual-interface SD and USB connections

- **Integrated Wi-Fi** — Eye-Fi produces several SD cards with built-in Wi-Fi transceivers providing static security (WEP 40; 104; and 128, WPA-PSK, and WPA2-PSK). The card lets any digital camera with an SD slot transmit captured images over a wireless network, or store the images on the card's memory until it is in range of a wireless network.[34] Some models geotag their pictures.
- **Pre-loaded content** — In 2006, SanDisk announced Gruvi, a microSD card with extra Digital Rights Management (DRM) features, with which to use memory cards as a medium to sell content. SanDisk again announced pre-loaded cards in 2008, under the slotMusic name, this time not using any of the DRM capabilities of the SD card.[35] In 2011, SanDisk offered various collections of 1000 songs on a single slotMusic card for about $40,[36] now restricted to compatible devices and without the ability to copy the files.

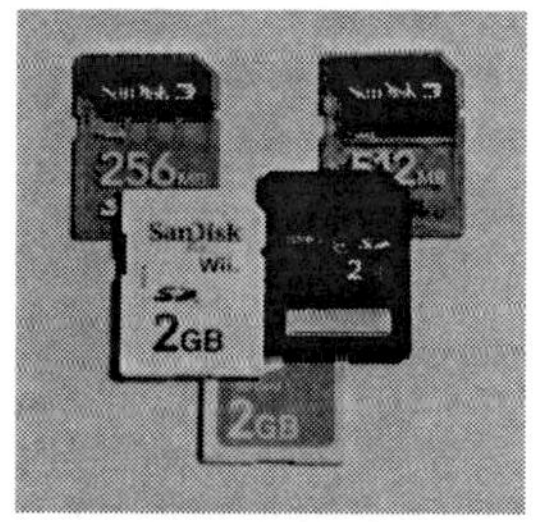

SanDisk cards in various colors

- **Integrated USB connector** — The SanDisk **SD Plus** product can be plugged directly into a USB port without needing a USB card reader.[37] Other companies introduced comparable products, such as the **Duo SD** product of OCZ Technology and the **3 Way** (microSDHC, SDHC, and USB) product of A-DATA, which was available in 2008 only.
- **Different colors** — SanDisk has used various colors of plastic or adhesive label, including a "gaming" line in translucent plastic colors that indicated the card's capacity.
- **Integrated display** — In 2006, A-DATA announced a **Super Info SD** card with a digital display that provided a two-character label and showed the amount of unused memory on the card.[38]

Speeds

An SD card's speed is measured by how quickly information can be read from, or written to, the card. In applications that require sustained write throughput, such as video recording, the device might not perform satisfactorily if the SD card's class rating falls below a particular speed. For example, a camcorder built for a Class 6 card may suffer dropouts or corrupted video if a slower card is used. Digital cameras may experience a noticeable lag between shots, while the camera writes the picture to a slower card.

A card's speed depends on many factors, such as the following:

- The likelihood of soft errors that the card's controller must re-try
- The fact that, on most cards, writing data requires the controller to read and erase a larger region, then rewrite that entire region with the desired part changed
- The possibility of fragmentation: that a body of information the host views as a unit is, for historical reasons, written to non-contiguous regions of memory. (This possibility does not cause rotational or head-movement delays as with magnetic media, but it does vary the amount of computation the card's controller must do.)

In early SD cards, the speed was measured with the "×" rating, which compared the average speed of reading data to that of the original CD-ROM drive. Currently, the official unit of measurement is the **Speed Class Rating**, which guarantees a minimum rate at which data can be written to the card.

The newer families of SD card improve card speed by increasing the bus rate (the frequency of the clock signal that strobes information into and out of the card). Whatever the bus rate, the card can signal to the host that it is "busy" until a read or a write operation is complete. Compliance with a higher speed rating is a guarantee that the card limits its use of the "busy" indication.

Speed Class Rating

The SD Association defines standard speed classes indicating minimum performance to record video. These are defined in terms of suitability for different applications:[39]

- Class 2 for SD video recording
- Class 4 and 6 for HD ~ Full HD video recording,
- Class 10 for Full HD video recording and HD still consecutive recording
- UHS Speed Class 1 for real-time broadcasts and large-size HD videos

The specification defines these classes in terms of performance curves which translate into the following minimum performance levels (on an empty card):[40]

32GB SDHC card

Class	Minimum performance
Class 2	2 MB/sec
Class 4	4 MB/sec
Class 6	6 MB/sec
Class 10	10 MB/sec

Speed Classes 2, 4, and 6 assert that the card supports the respective number of MB/sec as a minimum sustained write speed for a card in a fragmented state. Class 10 asserts that the card supports 10 MB/s as a minimum non-fragmented sequential write speed.[40] By comparison, the older "×" rating measured maximum speed under ideal conditions, and was vague as to whether this was read speed or write speed.

The host device can read a card's speed class, unlike the earlier "×" speed ratings. A device can warn the user if the card reports a speed class that falls below an application's minimum need.[40]

"×" rating

The "×" rating is a multiple of the standard CD-ROM drive speed of 150 kB/s (approximately 1.23 Mbit/s). Basic cards transfer data up to six times (6×) the CD-ROM speed; that is, 900 kB/s or 7.37 Mbit/s. The 2.0 specification defines speeds up to 200×, but is not as specific as Speed Classes are on how to measure speed. Manufacturers may report best-case speeds and may report the card's fastest read speed, which is typically faster than the write speed. Vendors including Transcend and Kingston report their cards' write speed.[41]

This table shows the approximate relationship between certain "×" ratings and the newer Speed Class designations:

Rating	Approx. (MB/s)	Comparable Speed Class
16×	2.34	2
32×	4.69	4
48×	7.03	6
100×	14.6	10

UHS Speed Class

The Ultra-High Speed (UHS) designation is available on some SDHC and SDXC cards.[42] The following ultra-high speeds are specified:

- **UHS-I** cards, specified in SD Version 3.01,[40] support a clock frequency of 100 MHz (a quadrupling of the original Default Speed), which in four-bit transfer mode could transfer 50 MB/s. UHS-I cards declared as **UHS104** also support a clock frequency of 208 MHz, which could transfer 104 MB/s. UHS-I is the only class for which products are currently available.[43]
- Double data rate operation at 50 MHz (DDR50) is also specified in Version 3.01, and is mandatory for microSDHC and microSDXC cards labeled as UHS-I. In this mode, four bits are transferred when the clock signal rises and another four bits when it falls, transferring an entire byte on each full clock cycle.
- **UHS-II** cards, to be defined in Version 4.0, further raise the data transfer rate to a theoretical maximum of 312 MB/s.[44][45]

UHS memory cards work best with UHS host devices. The combination lets the user record HD resolution videos to tapeless camcorders while performing other functions. It is also suitable for real-time broadcasts and capturing large HD videos.

Cards that comply with UHS show UHS-I or UHS-II on the label, and report this capability to the host device. Use of UHS requires that the host device command the card to drop from 3.3-volt to 1.8-volt operation and select the 4-bit transfer mode.

Market penetration

Secure Digital cards are used in many consumer electronic devices, and have become a widespread means of storing several gigabytes of data in a small size. Devices where the user may remove and replace cards often, such as digital cameras, camcorders, and video game consoles, tend to use full-sized cards. Devices where small size is paramount, such as mobile phones, tend to use microSD cards. SD cards are not the most economical solution in devices that need only a small amount of non-volatile memory, such as station presets in small radios. They may also not present the best choice for applications where higher storage capacities or speeds are a requirement as provided by other flash card standards such as CompactFlash.

microSD to SD adapter (left), microSD to miniSD adapter (middle), microSD card (right)

Many personal computers of all types and personal digital assistants (PDAs) use SD cards, either through built-in slots or through an active electronic adaptor. Adaptors exist for the PC card, ExpressBus, USB, FireWire, and the parallel printer port. Active adaptors also let SD cards be used in devices designed for other formats, such as CompactFlash. The FlashPath adaptor lets SD cards be used in a floppy disk drive.

Digital cameras

A camcorder with a 4 GB SDHC card

SD/MMC cards replaced Toshiba's SmartMedia as the dominant memory card format used in digital cameras. In 2001, SmartMedia had achieved nearly 50% use, but by 2005 SD/MMC had achieved over 40% of the digital camera market and SmartMedia's share had plummeted, with cards not being easily available in 2007.

At this time all the leading digital camera manufacturers use SD in their consumer product lines, including Canon, Casio, Fujifilm, Kodak, Leica, Nikon, Olympus, Panasonic, Pentax, Ricoh, Samsung, and Sony. Formerly, Olympus and Fujifilm used XD-Picture Cards (xD cards) exclusively, while Sony only used Memory Stick; however as of January 2010, all three support SD.

Some prosumer and professional digital camera models continue to offer CompactFlash, either on a second card slot or as the only storage, as they offer much higher capacities and faster transfer speeds and historically offered a better price/capacity ratio as well.

Secure Digital memory cards can be used in Sony XDCAM EX camcorders via the MEAD-SD01 adapter.[46]

Personal computers

USB-based universal Memory card reader

Although many personal computers accommodate SD cards as an auxiliary storage device through a built-in slot or a USB adaptor, SD cards cannot be used as the primary hard disk through the onboard ATA controller because none of the SD card variants supports ATA signalling. This use requires a separate SD controller chip[47] or a SD-to-CompactFlash converter. However, on computers that support bootstrapping from a USB interface, an SD card in a USB adaptor can be the primary hard disk, provided it contains an operating system that supports USB access once the bootstrap is complete.

Embedded systems

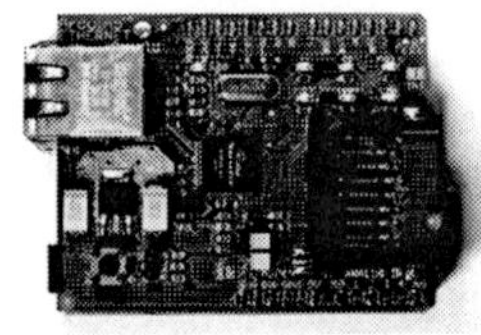
This shield (daughterboard) gives Arduino prototyping microprocessors access to SD cards plugged into the socket.

In 2008, the SDA specified Embedded SD, "leverag[ing] well-known SD standards" to enable non-removable SD-style devices on printed circuit boards.[48] SanDisk provides such memory components under the iNAND brand.[49]

Most modern microcontrollers have built-in SPI logic that can interface to a SD card operating in its SPI mode, providing non-volatile storage. Even if a microcontroller lacks the SPI feature, the feature can be emulated by bit banging. For example, a home-brew hack combines spare General Purpose Input/Output (GPIO) pins of the processor of the Linksys WRT54G router with MMC support code from the Linux kernel.[50] This technique can achieve throughput of up to 1.6 Mbit/s.

History

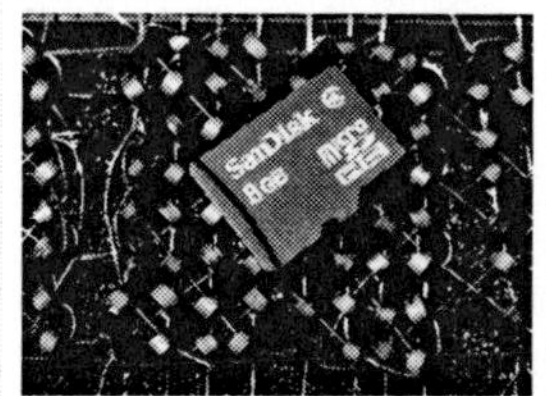
On top: a microSDHC card that stores about 8 billion bytes. Below it: 64 units of magnetic-core memory, used until the 1970s, that stored eight bytes.

In 1999, SanDisk, Matsushita, and Toshiba agreed to develop and market the Secure Digital (SD) Memory Card, which was a development of the MultiMediaCard (MMC). The new card provided both digital rights management (DRM) up to the Secure Digital Music Initiative (SDMI) standard, and a high memory density for the time.

The new format was designed to compete with the Memory Stick, a DRM product that Sony released the prior year. It was mistakenly predicted that DRM features[51] would be widely used due to pressure from music and other media suppliers to prevent piracy.

The trademarked *SD* logo was originally developed for the Super Density Disc, which was the unsuccessful Toshiba entry in the DVD format war. This is why the *D* resembles an optical disc.

At the 2000 Consumer Electronics Show (CES) trade show, the three companies announced the creation of the SD Card Association (SDA) to promote SD cards. The SD Card Association's headquarters are in San Ramon, California, United States and it comprises some 30 product manufacturers that make interoperable memory cards and devices. Early samples of the SD Card were available in the first quarter of 2000, with production quantities of 32 and 64 MB cards available 3 months later.

Relation to MMC

Size comparison of various flash cards: SD, CompactFlash, MMC, xD

The SD cards changed the MMC design in several ways:

- Asymmetrical slots in the sides of the SD card prevent inserting it upside down, while an MMC will go in most of the way but make no contact if inverted.
- Most SD cards are 2.1 mm thick, compared to 1.4 mm for MMCs. The SD specification defines a card called **Thin SD** with a thickness of 1.4 mm, but they are rare, as the SDA went on to define even smaller form factors.
- The card's electrical contacts are recessed beneath the surface of the card, protecting them from contact with a user's fingers.
- The SD specification envisaged capacities and transfer rates exceeding those of MMC, and these have both grown over time. For a comparison table, see below.

Full-sized SD cards will not fit into the slimmer MMC slots, and there are other issues that affect the ability to use one format in a host device designed for the other.

Follow-on products

In March 2003, SanDisk Corporation announced the introduction of the **miniSD** and demonstrated it at CeBIT 2003.[52] The SDA adopted the miniSD card in 2003 as a small form factor extension to the SD card standard. While the new cards were designed especially for use in mobile phones, they are usually packaged with a miniSD adapter which enables compatibility with all devices equipped with a standard SD memory card slot.

At CTIA Wireless 2005, the SDA announced the small **microSD** form factor (and SDHC, with capacities in excess of 2 GB and a minimum sustained read and write speed of 17.6 Mbit/s). SanDisk had conceived microSD when its CTO and the CTO of Motorola concluded that current memory cards were too large for mobile phones. The card was originally called T-Flash, but just before product launch, T-Mobile sent a cease and desist order to SanDisk claiming that T-Mobile owned the trademark on T-(anything), and the name was changed to TransFlash. TransFlash and microSD cards are the same; each can be used in devices made for the other.[53] SanDisk induced the SDA to administer the microSD standard. The SDA approved the final microSD specification on July 13, 2005. Initially, microSD cards were available in capacities of 32, 64, and 128 MB.

In April 2006, the SDA released a detailed specification for the non-security related parts of the SD memory card standard and for the Secure Digital Input Output (SDIO) cards and the standard SD host controller.

In September, 2006, SanDisk announced the 4 GB miniSDHC.[54] Like the SD and SDHC, the miniSDHC card has the same form factor as the older miniSD card but the HC card requires HC support built into the host device. Devices that support miniSDHC will work with miniSD and miniSDHC, but devices without specific support for miniSDHC will work only with the older miniSD card.

In January 2009, the SDA announced the SDXC family, which supports cards up to 2 TB and speeds up to 300 Mbyte/s.

Openness of the standard

Like most memory card formats, SD is covered by numerous patents and trademarks. Royalties for SD card licences are imposed for manufacture and sale of memory cards and host adapters (1000 USD/year plus membership at 1500 USD/year), but SDIO cards can be made without royalties.

Dismantled microSD to SD adapter showing the passive connection from the microSD card slot on the bottom to the SD pins on the top

Early versions of the SD specification were available only after agreeing to a non-disclosure agreement (NDA) that prohibited development of an open source driver. However, the system was eventually reverse-engineered, and free software drivers provided access to SD cards that did not use DRM. Since then, the (SDA) has provided a simplified version of the specification under a less restrictive license.[55] Although most open-source drivers were written before this, it has helped to solve compatibility issues.

In 2006, the SDA also released a simplified version of the specification of the host controller interface (as opposed to the specification of SD cards) and later also for the physical layer, ASSD extensions, SDIO, and SDIO Bluetooth Type-A, under a disclaimers agreement.[56] Again, most of the information had already been discovered and Linux had a fully free driver for it. Still, building a chip conforming to this specification caused the One Laptop per Child project to claim "the first truly Open Source SD implementation, with no need to obtain an SDI license or sign NDAs to create SD drivers or applications."[57]

The fact that the complete SD specification is proprietary mainly affects embedded systems and laptops, since users of desktop PCs generally read SD cards via USB-based card readers. These card readers present a standard USB mass storage interface to memory cards, thus separating the operating system from the details of the underlying SD interface. However, embedded systems (such as portable music players) usually gain direct access to SD cards and thus need complete programming information. Desktop card readers are themselves embedded systems; their manufacturers have usually paid the SDA for complete access to the SD specifications. Many notebook computers now include SD card readers not based on USB; device drivers for these essentially gain direct access to the SD card, as do embedded systems.

Technical details

Transfer modes

The physical interface comprises 9 pins, except that the miniSD card adds two unconnected pins in the center and the microSD card omits one of the two V_{SS} (Ground) pins.

SD card pin assignment	miniSD card pin assignment	microSD card pin assignment

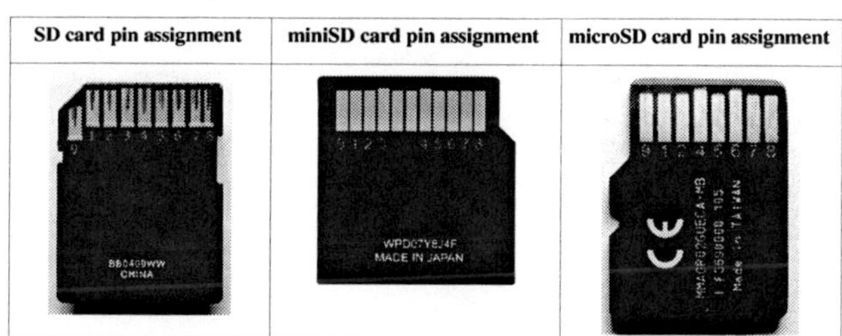

Various SD cards may support various combinations of the following bus types and transfer modes. The SPI bus and one bit SD bus are mandatory for all SD families, as explained in the next section.

- **SPI:** Serial Peripheral Interface Bus is primarily used by embedded microcontrollers. This bus type supports only a 3.3-volt interface.
- **One-bit SD:** Separate command and data channels and a proprietary transfer format.
- **Four-bit SD:** Uses extra pins plus some reassigned pins. UHS-I and UHS-II requires this bus type.

Once the host device and the SD card negotiate a bus interface, the usage of the numbered pins is the same for all three card sizes:

SPI Bus

Pin	Name	I/O	Logic	Description
1	nCS	I	PP	Card Select (Neg True)
2	DI	I	PP	Data In [MOSI]
3	VSS	S	S	Ground
4	VDD	S	S	Power
5	CLK	I	PP	Clock [SCLK]
6	VSS	S	S	Ground
7	DO	O	PP	Data Out [MISO]
8	NC nIRQ	. O	. OD	NC (Memory Cards) Interrupt (SDIO Cards)
9	NC	.	.	NC

One-Bit SD Bus

Pin	Name	I/O	Logic	Description
1	NC	.	.	NC
2	CMD	I/O	PP,OD	Command, Response
3	VSS	S	S	Ground
4	VDD	S	S	Power
5	CLK	I	PP	Clock
6	VSS	S	S	Ground
7	DAT0	I/O	PP	Data 0
8	NC nIRQ	. O	. OD	NC (Memory Cards) Interrupt (SDIO Cards)
9	NC	.	.	NC

Four-Bit SD Bus

Pin	Name	I/O	Logic	Description
1	DAT3	I/O	PP	Data 3
2	CMD	I/O	PP,OD	Command, Response
3	VSS	S	S	Ground
4	VDD	S	S	Power
5	CLK	I	PP	Clock
6	VSS	S	S	Ground
7	DAT0	I/O	PP	Data 0
8	DAT1 nIRQ	I/O O	PP OD	Data 1. SDIO Cards share with Interrupt Period
9	DAT2	I/O	PP	Data 2

Notes:

1. Direction is relative to card. I = Input, O = Output.
2. PP = Push-Pull logic, OD = Open-Drain logic.
3. S = Power Supply, NC = Not Connected (or logical high).

Interface

Command interface

SD cards and host devices initially communicate through a synchronous one-bit interface, where the host device provides a clock signal that strobes single bits into and out of the SD card. The host device thereby sends 48-bit commands and receives responses. The card can signal that a response will be delayed, but the host device can abort the dialogue.

Through issuing various commands, the host device can:

- Determine the type, memory capacity, and capabilities of the SD card
- Command the card to use a different voltage, different clock speed, or advanced electrical interface
- Prepare the card to receive a block to write to the flash memory, or read and reply with the contents of a specified block.

Inside a 512 MB SD card: NAND flash chip that holds the data (bottom) and SD controller (top)

The command interface is an extension of the MultiMediaCard (MMC) interface. SD cards dropped support for some of the commands in the MMC protocol, but added commands related to copy protection. By using only commands supported by both standards until determining the type of card inserted, a host device can accommodate both SD and MMC cards.

Inside a 2 GB SD card: two NAND flash chips (top and middle), SD controller chip (bottom)

Electrical interface

All SD card families initially use a 3.3-volt electrical interface. On command, SDHC and SDXC cards switch to 1.8-volt operation.[40]

At initial power-up or card insertion, the host device selects either the Serial Peripheral Interface (SPI) bus or the one-bit SD bus by the voltage level present on Pin 1. Thereafter, the host device may issue a command to switch to the four-bit SD bus interface, if the SD card supports it. For various card types, support for the four-bit SD bus is either optional or mandatory.[40]

After determining that the SD card supports it, the host device can also command the SD card to switch to a higher transfer speed. Until determining the card's capabilities, the host device should not use a clock speed faster than 400 kHz. SD cards other than SDIO (see below) have a Default Speed clock rate of 25 MHz. The host device is not required to use the maximum clock speed that the card supports. It may operate at less than the maximum clock speed to conserve power.[40] Between commands, the host device can stop the clock entirely.

Inside an SDHC card

SDIO cards

The SDIO family comprises Low-Speed and Full-Speed cards. Both types of SDIO cards support SPI and one-bit SD bus types. Low-Speed SDIO cards are allowed to also support the four-bit SD bus; Full-Speed SDIO cards are required to support the four-bit SD bus. To use a SDIO card as a "combo card" (for both memory and I/O), the host device must first select four-bit SD bus operation. Two other unique features of Low-Speed SDIO are a maximum clock rate of 400 kHz for all communications, and the use of Pin 8 as "interrupt" to try to initiate dialogue with the host device.[58]

Ganging cards together

The one-bit SD protocol was derived from the MMC protocol, which envisaged the ability to put up to 3 cards on a bus of common signal lines. The cards use open collector interfaces, where a card may pull a line to the low voltage level; the line is at the high voltage level (because of a pull-up resistor) if no card pulls it low. Though the cards shared clock and signal lines, each card had its own chip select line to sense that the host device had selected it.

The SD protocol envisaged the ability to gang 30 cards together without separate chip select lines. The host device would broadcast commands to all cards and identify the card to respond to the command using its unique serial number.

In practice, cards are rarely ganged together because open-collector operation has problems at high speeds and increases power consumption. Newer versions of the SD specification recommend separate lines to each card.

Achieving higher card speeds

The SD specification defines four-bit-wide transfers. (The MMC specification supports this and also defines an eight-bit-wide mode.) Transferring several bits on each clock pulse improves the card speed. Advanced SD families have also improved speed by offering faster clock frequencies and double data rate (explained here).

File system

Like other types of flash memory card, an SD card of any SD family is a block-addressable storage device, in which the host device can read or write fixed-size blocks by specifying their block number.

MBR and FAT

Most SD cards ship preformatted with one or more MBR partitions, where the first or only partition contains a file system. This lets them operate like the hard disk of a personal computer. Per the SD card specification, an SD card is formatted with MBR and the following file system:

- For SDSC cards: FAT16
- For SDHC cards: FAT32
- For SDXC cards: exFAT

Most consumer products that take an SD card will expect it to be partitioned and formatted in this way. The universal support for FAT16 and FAT32 allow the usage of SDSC and SDHC cards on most host computers with a compatible SD reader, to present the user with the familiar method of named files in a hierarchical directory tree.

On such SD cards, standard utility programs such as Mac OS X's "Disk Utility" or Windows' SCANDISK can be used to repair or retrieve corrupted data, and sometimes recover deleted files. Defragmentation tools for FAT file systems may be used on such cards. The resulting consolidation of files may provide a marginal improvement in the time required to read or write the file,[59] but not an improvement comparable to defragmentation of hard drives, where storing a file in multiple fragments may involve a time penalty to move between physical areas of the drive. Moreover, defragmentation performs writes to the SD card that count against the card's rated lifespan. The write endurance of the physical memory is discussed in the article on flash memory; newer technology to increase the storage capacity of a card currently provides worse write endurance.

When reformatting an SD card smaller than 4 GB, FAT16 should be used if the card is to be used in consumer devices. (FAT16 is also an option for 4 GB cards, but it requires the use of 64 kiB clusters, which are not widely supported.) FAT16 does not support cards above 4 GB.

The SDXC specification makes Microsoft's proprietary exFAT file system mandatory,[60] which is supported only by some proprietary operating systems.

Other file systems

Because the host views the SD card as a block storage device, the card does not require MBR partitions or any specific file system. The card can be reformatted to use any file system the operating system supports. For example:

- Under Unix-like operating systems such as Linux or FreeBSD, SD cards can be formatted using the UFS, Ext2, Ext3, Ext4, btrfs, HFS Plus, or ReiserFS file system.
- Under Mac OS X, SD cards can be partitioned as GUID devices and formatted with the HFS Plus file system.
- Under Windows and some Unix systems, SD cards can be formatted using NTFS and, on later versions, exFAT.

Any recent version of the above can format SD cards using the UDF file system.

Additionally, an SD card called Live SD can contain an embedded operating system (such as Live USB). Computers that can bootstrap from an SD card (either using a USB adapter or inserted into the computer's flash media reader) instead of the hard disk drive may thereby be able to recover from a corrupted hard disk drive. A Live SD can be write-locked to preserve the system's integrity.

Risks of reformatting

Reformatting an SD card with a different file system, or even with the same one, may make the card slower, or shorten its lifespan. Some cards use wear leveling, in which frequently modified blocks are mapped to different portions of memory at different times, and some wear-leveling algorithms are designed for the access patterns typical of the file allocation table on a FAT16 or FAT32 device.[61] In addition, the preformatted file system may use a cluster size that matches the erase region of the physical memory on the card; reformatting may change the cluster size and make writes less efficient. The SD Association provides free formatting software to overcome these problems.[62]

Power use

The power consumption of microSD cards varies by manufacturer, but appears to be in the range of 66–330 mW (20–100 mA at a supply voltage of 3.3 V). Specifications from TwinMos technologies list a maximum of 149 mW (45 mA) during transfer. Toshiba, on the other hand, lists 264–330 mW (80–100 mA).[63]

Storage capacity and incompatibilities

All SD cards let the host device determine how much information the card can hold, and the specification of each SD family gives the host device a guarantee of the maximum capacity a compliant card will report.

By the time the Version 2.0 (SDHC) specification was completed in June 2006,[64] vendors had already devised 2 GB and 4 GB SD cards, either as specified in Version 1.01, or by creatively reading Version 1.00. The resulting cards do not work correctly in some host devices.[65][66]

SDSC cards above 1 GB

A host device can ask any inserted SD card for its 128-bit identification string (the Card-Specific Data or CSD). In standard-capacity cards (SDSC), 12 bits identify the number of memory clusters (ranging from 1 to 4,096) and 3 bits identify the number of blocks per cluster (which decode to 4, 8, 16, 32, 64, 128, 256, or 512 blocks per cluster). The host device multiplies these figures (as shown in the following section) with the number of bytes per block to determine the card's capacity in bytes.

4 GB SDSC card

In SD version 1.00, the number of bytes per block was assumed to be 512. This permitted SDSC cards up to 4,096 × 512 × 512 = 1 GB, for which there are no known incompatibilities.

Version 1.01 let an SDSC card use a 4-bit field to indicate 1,024 or 2,048 bytes per block instead.[40] Doing so enabled cards with 2 GB and 4 GB capacity.

Early SDSC host devices that assume 512-byte blocks therefore will not fully support the insertion of 2 GB or 4 GB cards. In some cases, the host device can read data that happens to reside in the first 1 GB of the card. If the assumption is made in the driver software, success may be version-dependent. In addition, any host device might not support a 4 GB SDSC card, since the specification lets it assume that 2 GB is the maximum for these cards.

Storage capacity calculations

The format of the Card-Specific Data (CSD) register changed between version 1 (SDSC) and version 2.0 (which defines SDHC and SDXC).

Version 1

In Version 1 of the SD specification, capacities up to 2 GB are calculated by combining fields of the CSD as follows:

```
Capacity=(C_SIZE+1) × 2^(C_SIZE_MULT + 2 + READ_BL_LEN)
Where: 0 <= C_SIZE <= 4095,
       0 <= C_SIZE_MULT <= 7,
       READ_BL_LEN is 9 (for 512 bytes/sector) or 10 (for 1024 bytes/sector)
```

Later versions state (at Section 4.3.2) that a 2 GB SDSC card shall set its READ_BL_LEN (and WRITE_BL_LEN) to indicate 1024 bytes, so that the above computation correctly reports the card's capacity; but that, for consistency, the host device shall not request (by CMD16) block lengths over 512 bytes.[40]

Versions 2 and 3

In the definition of SDHC cards in Version 2.0, the C_SIZE portion of the CSD is 22 bits and it indicates the memory size in multiples of 512 KB. (The C_SIZE_MULT field is removed and READ_BL_LEN is no longer used to compute capacity.) Two bits that were formerly reserved now identify the card family: 0 is SDSC; 1 is SDHC or SDXC; 2 and 3 are reserved.[40] Because of these redefinitions, older host devices do not correctly identify SDHC or SDXC cards nor their correct capacity.

- SDHC cards are restricted to reporting a capacity not over 32 GiB.
- SDXC cards are allowed to use all 22 bits of the C_SIZE field. An SDHC card that did so (reported C_SIZE > 65375 to indicate a capacity of over 32 GiB) would violate the specification. A host device that relied on C_SIZE rather than the specification to determine the card's maximum capacity might support such a card, but the card might fail in other SDHC-compatible host devices.

Capacity is calculated thus:

```
Capacity=(C_SIZE+1)*524288
where for SDHC        4112<=C_SIZE<=65375    (approx. 2 GB) < capacity < 32 GiB
      for SDXC       65535<=C_SIZE                   32 GiB <= capacity <= 2 TiB max.
```

Capacities above 4 GB can only be achieved by following Version 2.0 or later versions. In addition, capacities equal to 4 GB must also do so to guarantee compatibility.

Comparison to other flash memory formats

Overall, SD is less open than CompactFlash or USB flash memory drives; these are open standards which can be implemented free of payment for licensing, royalties, or documentation. (CompactFlash and USB flash drives may, however, require licensing fees for the use of the SDA's trademarked logos.)

However, SD is much more open than Memory Stick, for which no public documentation nor any documented legacy implementation is available. All SD cards can be accessed freely using the well-documented SPI bus.

xD cards are simply 18-pin NAND flash chips in a special package and support the standard command set for raw NAND flash access. Although the raw hardware interface to xD cards is well understood, the layout of its memory contents—necessary for interoperability with xD card readers and digital cameras—is totally undocumented. The consortium that licenses xD cards has not released any technical information to the public.

Type	MMC	RS-MMC	MMC Plus	SecureMMC	SDIO	SD	miniSD	microSD
SD Socket	Yes	Extender	Yes	Yes	Yes	Yes	Adapter	Adapter
Pins	7	7	13	7	9	9	11	8
Width	24 mm	24 mm	24 mm	24 mm	24 mm	24 mm	20 mm	11 mm
Length	32 mm	18 mm	32 mm	32 mm	32 mm+	32 mm	21.5 mm	15 mm
Thickness	1.4 mm	1.4 mm	1.4 mm	1.4 mm	2.1 mm	2.1 mm (most) 1.4 mm (rare)	1.4 mm	1 mm
SPI mode	Optional	Optional	Optional	Yes	Yes	Yes	Yes	Yes
1-bit mode	Yes	Yes	Yes	Yes	Yes	Yes	Yes	Yes
4-bit mode	No	No	Yes	No	Optional	Optional	Optional	Optional
8-bit mode	No	No	Yes	No	No	No	No	No
Interrupts	No	No	No	No	Optional	No	No	No
Max clock rate	20 MHz	20 MHz	52 MHz	20 MHz?	50 MHz	208 MHz	208 MHz	208 MHz
Max transfer	20 Mbit/s	20 Mbit/s	416 Mbit/s	20 Mbit/s?	200 Mbit/s	832 Mbit/s	832 Mbit/s	832 Mbit/s
Max SPI transfer	20 Mbit/s	20 Mbit/s	52 Mbit/s	20 Mbit/s	50 Mbit/s	50 Mbit/s	50 Mbit/s	50 Mbit/s
DRM	No	No	No	Yes	N/A	Yes	Yes	Yes
User encrypt	No	No	No	Yes	No	No	No	No
Simplified spec	Yes	Yes	No	Not yet?	Yes	Yes	No	No
Membership cost	JEDEC: USD $4400/yr, optional				SD Card Association: USD $2000/yr, general; USD $4500/yr, executive			
Specification cost	Free			?	Simplified spec: Free Full spec: Free to members, USD $1000/yr to R&D non-members			
Host license	No	No	No	No	Yes, USD $1000/yr. No, if using SPI-mode only.			
Card royalties	Yes	Yes	Yes	Yes	Yes, $1000/yr	Yes	Yes	Yes
Open source compatible	Yes	Yes	Yes?	Yes?	Yes	Yes	Yes	Yes

Nominal operating voltage	3.3V	1.8V/3.3V	1.8V/3.3V[67][68]	1.8V/3.3V	3.3V	3.3V (SDSC), 1.8/3.3V (SDHC), 1.8/3.3V (SDXC)	3.3V (miniSD), 1.8/3.3V (miniSDHC)	3.3V (SDSC), 1.8/3.3V (microSDHC), 1.8/3.3V (microSDXC)
Maximum available capacity	128GB	2GB	128GB?	128GB?	?	4GB (SD), 32GB (SDHC), 128GB (SDXC)	4GB (miniSD), 16GB (miniSDHC)	4GB (microSD), 32GB (microSDHC), 64GB (microSDXC)
Type	**MMC**	**RS-MMC**	**MMC Plus**	**SecureMMC**	**SDIO**	**SD**	**miniSD**	**microSD**

- Table data compiled from MMC, SD, and SDIO specifications from SD Association and JEDEC web sites. Data for other card variations are interpolated.
- Capacity limit in all SD/MMC formats appears to be 128 GB in LBA mode (28-bit sector address).

References

[1] "About" (http://www.sdcard.org/developers/about/). SD Card Association. . Retrieved 2011-12-08.

[2] "Capacity" (http://www.sdcard.org/developers/overview/capacity/). SD Card Association. . Retrieved 2011-12-08.

[3] "Using SDXC" (http://www.sdcard.org/consumers/sdxc_capabilities/using_sdxc/). SD Card Association. . Retrieved 2011-12-08.

[4] "SDIO" (http://www.sdcard.org/developers/overview/sdio/). SD Card Association. . Retrieved 2011-12-08.

[5] "Using SD Memory Cards is Easy" (video on YouTube) – SD Card Association (http://www.youtube.com/watch?v=wX6NwBa1csY)

[6] "About" (https://www.sdcard.org/consumers/choices/). SD Card Association. . Retrieved 2011-05-02.

[7] What are SDHC, miniSDHC, and microSDHC? (http://www.sandisk.com/Assets/File/pdf/retail/SDHC1.pdf) SanDisk.com

[8] What's new in Firmware 2.41 Beta (for COWON D2) (http://www.jetaudio.com/download/cowon_rn_d2.html#241) JetAudio

[9] "Microsoft Support 934428 – Hotfix for Windows XP that adds support for SDHC cards that have a capacity of more than 4 GB" (http://support.microsoft.com/kb/934428). Support.microsoft.com. 2008-02-15. . Retrieved 2010-08-22.

[10] "Microsoft Support 939772 – Some Secure Digital (SD) cards may not be recognized in Windows Vista" (http://support.microsoft.com/kb/939772). Support.microsoft.com. 2008-05-15. . Retrieved 2010-08-22.

[11] "Microsoft Support 949126 – A Secure Digital High Capacity (SDHC) card is not recognized on a Windows Vista Service Pack 1-based computer" (http://support.microsoft.com/kb/949126). Support.microsoft.com. 2008-02-21. . Retrieved 2010-08-22.

[12] "SanDisk and Sony to expand Memory Stick Pro and Memory Stick Micro formats" (http://www.sandisk.com/about-sandisk/press-room/press-releases/2009/2009-01-07-sandisk-and-sony-to-expand-âmemory-stick-proâ-and-âmemory-stick-microâ-formats). Sandisk.com. . Retrieved 2010-08-22.

[13] "SD Card, Memory Stick formats to reach 2 terabytes, but when? – Betanews" (http://www.betanews.com/article/SD_Card_Memory_Stick_formats_to_reach_2_terabytes_but_when/1231453659). . 090108 betanews.com

[14] "Pretec introduces world's first SDXC card: Digital Photography Review" (http://www.dpreview.com/news/0903/09030601pretecsdxc.asp). Dpreview.com. 2009-03-06. . Retrieved 2010-08-22.

[15] "Canon EOS Rebel T2i / 550D Digital SLR Camera Review" (http://www.the-digital-picture.com/Press-Release/Canon-EOS-Rebel-T2i-550D-Digital-SLR-Camera-Press-Release.aspx). . The-Digital-Picture.com

[16] Ng, Jansen (2009-11-24). "Lack of Card Readers Holding Back SDXC Flash Memory Adoption" (http://www.dailytech.com/Lack+of+Card+Readers+Holding+Back+SDXC+Flash+Memory+Adoption/article16915.htm). DailyTech. . Retrieved 2009-12-22.

[17] Ng, Jansen (2009-11-30). "Lenovo, HP, Dell Integrating SDXC Readers in New 32nm Intel "Arrandale" Laptops" (http://www.dailytech.com/Lenovo+HP+Dell+Integrating+SDXC+Readers+in+New+32nm+Intel+Arrandale+Laptops/article16937.htm). DailyTech. . Retrieved 2009-12-22.

[18] Ng, Jansen (2009-12-22). "Toshiba Sampling First SDXC Flash Memory Cards" (http://www.dailytech.com/Toshiba+Sampling+First+SDXC+Flash+Memory+Cards/article16972.htm). DailyTech. . Retrieved 2009-12-22.

[19] "Toshiba's 64 GB SDXC card to finally go on sale (in Japan)" (http://www.crunchgear.com/2010/04/15/toshibas-64gb-sdxc-card-to-finally-go-on-sale-in-japan/). CrunchGear. . Retrieved 2010-08-09.

[20] "Panasonic Introduces New 64 GB* and 48 GB* SDXC Memory Cards, Available Globally in February 2010" (http://www2.panasonic.com/webapp/wcs/stores/servlet/prModelDetail?storeId=11301&catalogId=13251&itemId=389511&modelNo=Content01052010041118461&surfModel=Content01052010041118461). Panasonic. . Retrieved 2010-08-09.

[21] "Sandisk ships its highest capacity sd card ever" (http://sandisk.com/about-sandisk/press-room/press-releases/2010/2010-02-22-sandisk-ships-its-highest-capacity-sd-card-ever,-the-64gb-sandisk-ultra-sdxc-card). SanDisk. . Retrieved 2010-08-09.

[22] Lexar ships 128 GB Class 10 SDXC card; March 2011. (http://www.betanews.com/article/Lexar-ships-first-128GB-SDXC-cards/1300305310)

[23] "SDXC/SDHC 433X Class 16 Card from Pretec" (http://www.pretec.com/news-event/press-room/item/root/sdxcsdhc-433x-c16). Pretec. 2011-06-13. . Retrieved 2010-12-03.

[24] First 64GB microSD Card Here; When Will Smartphones Support It? (http://pocketnow.com/smartphone-news/first-64gb-microsd-card-arrives-when-will-smartphones-support)

[25] Kingmax flaunts world's first 64GB microSD card (http://www.engadget.com/2011/05/26/kingmax-flaunts-worlds-first-64gb-microsd-card/)

[26] About Compatibility with Host Devices (https://www.sdcard.org/consumers/compatibility/) SD Association

[27] WindowsXP-KB955704-x86-ENU.exe (2009-01-26). "Download: Update for Windows XP (KB955704) – Microsoft Download Center – Download Details" (http://www.microsoft.com/downloads/details.aspx?FamilyID=1cbe3906-ddd1-4ca2-b727-c2dff5e30f61&displaylang=en). Microsoft.com. . Retrieved 2012-03-21.

[28] "About the SD and SDXC card slots" (http://support.apple.com/kb/HT3553). Apple Inc.. 2011-05-03. . Retrieved 2011-09-05.

[29] "Apple released exFAT support in OS X 10.6.5 update" (http://www.tuxera.com/mac/apple-released-exfat-support-in-os-x-10-6-5-update/). Tuxera.com. 2010-11-22. . Retrieved 2012-01-04.

[30] Exfat for BSD and Linux systems from Google Code (http://code.google.com/p/exfat/)

[31] "Windows Phone 7 Secure Digital Card Limitations" (http://support.microsoft.com/kb/2450831). .

[32] "Windows Phone 7's microSD mess: the full story (and how Nokia can help you out of it)" (http://www.engadget.com/2010/11/17/windows-phone-7s-microsd-mess-the-full-story-and-how-nokia-ca). .

[33] "Super Talent Super Digital Overview & Specs" (http://reviews.cnet.com/secure-digital/super-talent-super-digital/4505-3236_7-32074490.html). Cnet.com. . Retrieved 2012-07-28.

[34] "Home" (http://www.eye.fi). Eye-Fi. . Retrieved 2010-08-22.

[35] AudioHolics (http://www.audioholics.com/news/industry-news/sandisk-slotmusic)

[36] "slotRadio" (http://www.sandisk.com/consumer-products/slotradio). SanDisk. . Retrieved 2011-11-27.

[37] SanDisk Ultra II SD Plus USB/SD card (http://www.theregister.co.uk/2005/07/25/review_sandisk_ultra_ii_sd_plus/) The Register, 2005-07-25

[38] "A-DATA Super Info SD Card 512MB" (http://www.techpowerup.com/reviews/AData/ADATASuperInfoSD/). techpowerup.com. 2007-02-20. . Retrieved 2011-12-30.

[39] SD Speed Class (https://www.sdcard.org/consumers/speed/) SDCard.org

[40] "SD Part 1, Physical Layer Simplified Specification, Version 3.01" (http://www.sdcard.org/downloads/pls/simplified_specs/Part_1_Physical_Layer_Simplified_Specification_Ver_3.01_Final_100518.pdf). . Retrieved 2011-12-04.

[41] "Kingston Technology Company – Flash Memory Cards and X-Speed Ratings" (http://www.kingston.com/flash/x/default.asp). Kingston.com. . Retrieved 2010-08-22.

[42] "SD cards branded with an upper-case 'I' are faster, yo" (http://www.engadget.com/2010/06/24/sd-cards-branded-with-an-upper-case-i-are-faster-yo/). Engadget. . Retrieved 2010-08-22.

[43] "SD Speed Class/UHS Speed Class" (https://www.sdcard.org/developers/overview/speed_class/). . Retrieved 21 Nov 2011.

[44] "SD Card Association Triples Speeds with UHS-II" (https://www.sdcard.org/press/SD_Association_Announces_UHS-II_eBOOK_Jan_5_2011_ENGLISH.PDF). 5 January 2011. . Retrieved 2011-08-09.

[45] "SD Card Association announces UHS-II, ultra high-speed SD card specification" (http://www.robgalbraith.com/bins/content_page.asp?cid=7-11133-11156). Rob Galbraith. 2011-01-05. . Retrieved 2011-01-05.

[46] MEAD-SD01 SDHC card adapter (Sony) (http://pro.sony.com/bbsc/ssr/micro-xdcamexsite/cat-accessories/product-MEADSD01/)

[47] "TS-7800 Embedded" (http://www.embeddedarm.com/products/board-detail.php?product=TS-7800). Embeddedarm.com. . Retrieved 2010-08-22.

[48] "Embedded SD" (http://www.sdcard.org/developers/overview/embedded_sd/). SD Card Association. . Retrieved 2011-11-30.

[49] "iNAND Embedded Flash Drives" (http://www.sandisk.com/business-solutions/inand-embedded-flash-drives). SanDisk. . Retrieved 2011-11-30.

[50] "Linksys WRT54G-TM SD/MMC mod – DD-WRT Wiki" (http://www.dd-wrt.com/wiki/index.php/Linksys_WRT54G-TM_SD/MMC_mod). Dd-wrt.com. 2010-02-22. . Retrieved 2010-08-22.

[51] "Press Releases 17 July 2003" (http://www.toshiba.co.jp/about/press/2003_07/pr1701.htm). Toshiba. 2003-07-17. . Retrieved 2010-08-22.

[52] SanDisk Introduces The World's Smallest Removable Flash Card For Mobile Phones-The miniSD Card (http://www.sandisk.com/Corporate/PressRoom/PressReleases/PressRelease.aspx?ID=1536) SanDisk.com

[53] PhoneScoop – Sandisk T-Flash announcement. (http://www.phonescoop.com/news/item.php?n=801)

[54] SanDisk Introduces 4GB miniSDHC Flash Card for Mobile Phones (http://www.sandisk.com/Corporate/PressRoom/PressReleases/PressRelease.aspx?ID=3530) SanDisk.com

[55] "Sharp Linux PDA promotes the use of proprietary SD card, but more open MMC works just fine" (http://www.linux.com/archive/feed/20060). Linux.com. . Retrieved 2010-08-22.

[56] Simplified Specification Agreement (http://www.sdcard.org/developers/tech/sdcard/pls/) from the SDA's website
[57] "OLPC mailing list archive" (http://mailman.laptop.org/pipermail/community-news/2006-September/000023.html). Mailman.laptop.org. . Retrieved 2010-08-22.
[58] "Simplified Version of SDIO CARD SPEC" (https://www.sdcard.org/developers/overview/sdio/sdio_spec/). SD Card Association. . Retrieved 2011-12-09.
[59] Fragmentation and Speed (https://www.sdcard.org/developers/overview/speed_class/) SDCard.org
[60] "SDXC memory cards promise 2TB of storage, 300MBps transfer" (http://www.engadget.com/2009/01/07/sdxc-memory-cards-promise-2tb-of-storage-300mbps-transfer/). Engadget.com. . Retrieved 2010-08-22.
[61] "Optimizing Linux with cheap flash drives" (http://lwn.net/Articles/428584/). Linux Weekly News. . Retrieved 2011-04-11.
[62] SD Formatter 3.1 for SD/SDHC/SDXC (https://www.sdcard.org/downloads/formatter_3/), SD Association
[63] microSD & microSDHC Cards (http://www.toshiba-memory.com/en/micro_sd_cards.html) TOSHIBA Memory Solutions
[64] A look into how SDHC will affect the future Nand Flash market (http://www.dramexchange.com/WeeklyResearch/Post/1/492.aspx). DRAMeXchange, December 2006
[65] SD Compatibility (http://www.hjreggel.net/cardspeed/special-sd.html), CARDSPEED – Card Readers and Memory Cards, December 1, 2006
[66] "WinXP SP3 can't read 4GB SD card in multicard reader" (http://www.eggheadcafe.com/software/aspnet/33338344/winxp-sp3-cant-read-4gb.aspx). Eggheadcafe.com. . Retrieved 2010-08-22.
[67] JEDEC MMC 4.4 Standard Pg.7 (http://www.jedec.org/download/search/JESD84-A44.pdf), http://www.jedec.org 2008
[68] Transcend v4.0 card does not support 1.8V (http://www.transcendusa.com/support/dlcenter/datasheet/TSxxMMC4.pdf), http://www.transcendusa.com 2009

External links

Organizations

- Official website (http://www.sdcard.org), SD Card Association, sdcard.org
  - Membership: $2000/yr for General, $4500/yr for Executive.
  - Full Specification: Free for members, $1000/yr for R&D non-members.

Specifications

- SD Simplified Specifications (https://www.sdcard.org/downloads/pls/), sdcard.org, Free
  - Part 1, Physical Layer, v3.01
  - Part A1, ASSD Extension, v2.00
  - Part A2, SD Host Controller, v3.00
  - Part E1, SDIO, v3.00
  - Part E2, SDIO Bluetooth Type-A, v1.00
- Microsoft Extensible Firmware Initiative FAT32 File System Specification (http://download.microsoft.com/download/1/6/1/161ba512-40e2-4cc9-843a-923143f3456c/fatgen103.doc), Microsoft

Software

- SD Formatter for SD / SDHC / SDXC cards (Windows and Mac) (https://www.sdcard.org/downloads/formatter_3/), sdcard.org

Comparisons

- Comparison of numerous memory cards and readers, plus technical information (http://www.hjreggel.net/cardspeed/index.html#special-sd.html), hjreggel.net
- SDHC Card comparison (german) (http://www.hardware-infos.com/tests.php?test=81), hardware-infos.com
- Comparison list of common SDHC/SDXC cards (http://www.valuetech.de/bestenliste/sdhc), valuetech.de
- Speed Testing of UHS-1 Cards in prosumer Nikon D7000 Camera (http://glamourphotography.co/gear/uhs-speed-class-1-sdhc-memory-cards-tested-delkin-8gb-elite-633x-secure-digital-uhs-i-95-mbsec-vs-sandisk-extreme-p), Glamour Photography
- Nikon D7000 Memory speed tests, SDHC Memory Cards: including Sandisk Extreme Pro UHS Speed Class 1 45 Mbyte/s and Several Popular Class 10 SDHC Cards (http://glamourphotography.co/gear/nikon-d7000-memory-speed-tests-sdhc-memory-cards-including-sandisk-extreme-pro-uhs-speed-class-1-45mbsec-and-s

), Glamour Photography

Interfacing

- Interfacing to SD cards, great technical details (http://elm-chan.org/docs/mmc/mmc_e.html)
- Interfacing AVR (Arduino) to SD cards, C source code (http://www.dharmanitech.com/2009/01/sd-card-interfacing-with-atmega8-fat32.html)
- Interfacing ARM to SD cards, C source code (http://gandalf.arubi.uni-kl.de/avr_projects/arm_projects/arm_memcards/index.html)
- Interfacing MSP430 to SD cards, C source code (http://www.cs.ucr.edu/~amitra/sdcard/Additional/sdcard_appnote_foust.pdf), Michigan State University
- Interfacing MAXQ2000 to SD cards, good technical descriptions, C source code (http://www.maxim-ic.com/appnotes.cfm/an_pk/3969), maxim-ic.com
- SD card controller, Verilog source code (http://www.opencores.org/project,sdcard_mass_storage_controller), opencores.org

Article Sources and Contributors

Nokia_X2-01 *Source*: http://en.wikipedia.org/w/index.php?title=Nokia_X2-01 *Contributors*: Bunnyhop11, CaroleHenson, Debarshisharma, Gene93k, Hmains, Interungulate, JamesBWatson, LittleWink, Pakshya, Scientist Haley, Sidthatsme, 4 anonymous edits

Feature_phone *Source*: http://en.wikipedia.org/w/index.php?title=Feature_phone *Contributors*: Acalamari, Andries, Aqwis, Chenopodiaceous, ChrisUK, CoolingGibbon, Davidbengtenglund, Earle Martin, Gamliel Fishkin, Giraffedata, Gnulinux, Gordon Ecker, HalfShadow, Imroy, JNW, Jason.grossman, Lester, Magicoast, Mange01, Mdwh, MonkeyKingBar, NeOFreedom, Obiwankenobi, Pgallert, Silverk331, Syp, Thumperward, Whoop whoop pull up, Yetanotherkontributor, □ , 57 anonymous edits

Series_40 *Source*: http://en.wikipedia.org/w/index.php?title=Series_40 *Contributors*: 1exec1, Agreimann, Alex Perry, Aniac, Ashrutisingh, Azuya, Bjelleklang, Blackvienna, Blakegripling ph, C933103, Calltech, Canaima, Centrx, Chapzboy, Charivari, Cnj, Daverocks, Davidbengtenglund, Diwas, Djlarz, Djr xi, Dktz, Dublinclontarf, Elspif, Ettrig, Evilboy, FleetCommand, Flod logic, Florin92, GRighta, Gaius Cornelius, Gaming&Computing, Hjorten, Imroy, ItsMeOrYou, JLaTondre, Jarekadam, Jasonon, Jesjimher, Joneilim, LordBen5000, MMuzammils, Mburdis, Mdwh, Milan292208, Monkeyhousetim, Msulik, Munamankeli, Park3r, Petri Krohn, Polisher of Cobwebs, Polluks, Rich Farmbrough, Rlobkovsky, Roleplayer, Rudiedude, Rwwww, SimonMenashy, Sofoklas, The Seventh Taylor, TheParanoidOne, Tirkfl, Txuspe, UKER, Vegaswikian, W like wiki, We hope, Xkury, Zakawer, ZeroOne, Zhernovoi, 106 anonymous edits

Nokia_phone_series *Source*: http://en.wikipedia.org/w/index.php?title=Nokia_phone_series *Contributors*: ARMSidewinder, Agent L, Aishwaryarajsahai, Andries, BD2412, Beyoncé Superfanatic, Bmusician, Ceelo99, Cnj, CommonsDelinker, DiscoverYellow, Editor182, Eik Corell, Feezo, FinnsDeal, Fæ, GoingBatty, Havarhen, ItsMeOrYou, JaGa, Jason24589, Jerryobject, Mardus, Maxinbjohn, Maxxyme, N90p, Ohconfucius, R'n'B, SchreiberBike, StAnselm, Tbhotch, They, Ushio01, WayKurat, Whoop whoop pull up, Woohookitty, Zunmun, 39 anonymous edits

Mobile_operating_system *Source*: http://en.wikipedia.org/w/index.php?title=Mobile_operating_system *Contributors*: 0v192, 12Daanny123, 9point6, A Quest For Knowledge, A tumiwa, Adileader, AgarwalSumeet, Akriesch, Aldwaik, Aleph Infinity, Alunphillips, Ammon86, Andries, Anna Lincoln, Aravindan Shanmugasundaram, Aurora317, Bender235, Bgkwtnyqhzor, Binoyjsdk, Bomazi, Bongwarrior, Bounce1337, C933103, Cab88, CalumCook234, Caulde, Cdw1952, Cenora, Chmyr, Chris the speller, Colonies Chris, DKqwerty, Darkguy, DaveChild, Dawnseeker2000, Derek R Bullamore, Diego Moya, DominiqueHazaelMassieux, Download, Drench, Efa, Egil, Enterprise12, Eraserhead1, Eried, Erimaxbau, Evan-Amos, EvenGreenerFish, Fatka, FeatherPluma, Ferritecore, Ffooxx 2006, Fiftyquid, Frap, Frederik Blem Vigold, Gabriel2008, Ghettoblaster, Gnaphosa, GoingBatty, Ham1000, Hammersoft, Hempanicker, Hopbit, Hypon888, IBoy2G, Illegal Operation, Immunmotbluescreen, Interframe, InternetMeme, ItsMeOrYou, Iuhkjhk87y678, Jack007, Jaz246, Jerryobject, John of Reading, Keith139, Kiore, Kotiyanimr, Kozuch, Lester, LifeH2O, Lightmouse, LilHelpa, Lprd2007, Lumialover, Lun Esex, MA LópezMolina, MER-C, Macungie, Mancini, Manuel.nas, Mark Arsten, Materialscientist, Mc nacho, Mdwh, Megarhythm, Melab-1, Melmann, Mike Restivo, Mikeormike, Milominderbinder2, Mjdtjm, Morning Sunshine, Mr left, N2e, Nekohan, Nemesis of Reason, NerdyScienceDude, Netemcee, Niceguyedc, Nono64, NotWith, Oicur0t, OliverHector, Onyxqk, Oscaroe, Over shown badn, PXUmais, Paraplegicemu, PieterDeBruijn, Quake44, R'n'B, RP.CS11, Reaper Eternal, Red Act, Redsparta, Rich Farmbrough, Rogernightman, Rothgar, Rreagan007, Rwwww, Rzęsor, SF007, Sam4others, Samjack91, Shankarnikhil88, Skierpage, Smartmo, Solanki.3108, Solphusion, Sonixrulerz, Spidermario, Stekse, Stephans88, Stichbury, SudoGhost, Tahir mq, Taylortbb, The Anome, The RedBurn, TheWikiAuthor, Themfromspace, Thewikimonkey, Thumperward, True Tech Talk Time, Ukexpat, Useerup, Viking90, Virtualerian, VoluntarySlave, Wamp wamp, Wasbeer, Wave4, Wikitürkçe, WinampLlama, Woohookitty, Wysprgr2005, Xcvista, Yorxs, YuMaNuMa, Zarinfam, Zayani, Zundark, 827 anonymous edits

Nokia *Source*: http://en.wikipedia.org/w/index.php?title=Nokia *Contributors*: -Majestic-, 007zoo, 130.236.221.xxx, 159753, 16@r, 1exec1, 205ywmpq, 28421u2232nfenfcenc, A bit iffy, A box of sticks, A. B., A333, AB, ABF, AEMoreira042281, Abdul raja, Abnyc81, Academic Challenger, Accurate Nuanced Clear, Acdx, Acela Express, Adamrush, Adaobi, Adashiel, Adraeus, Aesopos, Aeusoes1, Agusm266, Ahazred8, Ahmedcena, Ahoerstemeier, Aijoovai, Aintneo, Alansohn, Aldis90, Alepik, Alex43223, Alexius08, AlfredWalsh, AliShaikh85, Alirezazzz, Alisha.4m, Allanvs, Allpower, Amitn, Andres, AndrewHowse, Andros 1337, Andyabides, Anir1uph, Anole 418, AnonMoos, Anshuln95, Antandrus, Antoncampos, Anubhavsharmaa, Apalsola, ApnAEA, Apparition11, Arch dude, Ariele, Armando, Artem-S-Tashkinov, Arthena, Arungm29, Arunsingh16, Asanka000, Ashmoo, Ashutosh.manager007, Atanasov, Avoided, Axeman89, BLAZEXBOY, BaSH PR0MPT, Backwalker, Badr55, Barek, Batbayarl, Beaker1306, Beao, Bearcat, Beetstra, Bender235, Benhocking, BernardZ, Bfaabaa, Big Brother 1984, BigT333, Billyboy1980, Blahma, Blake-, Blakegripling ph, Blobglob, Bloggert, Bluezy, Bmannaa, Bobblewik, Bobo192, Bogdangiusca, Bongomatic, Bongwarrior, Boomshadow, BorgHunter, BorisxXx, BoyanSyarov, Brennish, BrightStarSky, Brockert, BrokenSegue, Brokestudent007, Bruce1ee, BunnyT, C. A. Russell, C.Fred, C311u1ar, C628, CONFIQ, Cablecord, Cameron Scott, Can't sleep, clown will eat me, CanadianLinuxUser, Canterbury Tail, Capricorn42, Catiana 465, CecilWard, Cerebrith, Chamal N, Cheezy man, Cherkash, Cheung1303, Choptube, Chris the speller, ChrisHodgesUK, Chrisissocool, Chronulator, Clarkedexter, ClementSeveillac, CliffC, Clipmode, Closedmouth, Cmreditor, Cntras, Codey123, Coguar, Colibri37, Conversion script, Coolbull20, Coolslko, Cpl Syx, Crevox, Cryonic07, Crysb, Cybercobra, DBigXray, Dale Arnett, Damiens.rf, Damirgraffiti, Dancter, Daniel*D, Daniel575, DanielCD, Danim, Danio, DarkSaber2k, Darth Panda, Davidbspalding, Dazman 1988, Dck7777, Dearsina, Debresser, Delta avi delta, Den fjättrade ankan, Desaivishal14, Desbiadi, Diasimon2003, Dicklyon, Dicostathomas, Dima1, Discospinster, Dissolve, Diyar se, Djr xi, DmitTrix, DocendoDiscimus, Doctorcasey, DomQ, Dostal, Dostick, Download, Drewt, Drpickem, Dsavi.x4, Dyl, E Pluribus Anthony, E000xm, ES Vic, EWikist, Ed g2s, Editor182, Edsuom, Educatednawab, Edward, Eekerz, Eenu, Elfguy, Eliz81, Emmalewis1, Enemenemu, Epoxed, Eraserhead1, Erunestian, Esebi95, Estoy Aquí, Eta 94, Etincelles, Eurocanna, Everyking, FR Soliloquy, Falcon8765, Falconoffrance63, Faramir1138, Feezo, Felix Dance, Fffaaattt, Fieldday-sunday, Fixer88, FleetCommand, Flewis, Flora, Florentino floro, Florin92, Flrn, Flyguy649, Folksong, Fooishbar, Force39, Fraggle81, Fram, France64160, Franz-kafka, Fredrik, FreplySpang, Frymaster, FunkyDuffy, GMRE, Gachen, Galeon54, Galoubet, Gambit 28, Gandhietami, Gareth Griffith-Jones, Gaurav13dubey, Gavinio, Gdo01, Georgy90, Gethresh, Ghodannywahyudi99, Gimboid13, Gnangarra, Gnuton, Gobonobo, Godospoons, Gogo Dodo, Golemeye, Gr1st, Grafen, Graham87, Gratom, GreenJellybean, GreenJoe, Greenshed, Gregorydavid, Gringer, Groshna, Gsarwa, Guaka, Gump Stump, Gunglewack, Gunmetal Angel, Guy M, H.lloyd, H0dd0ck, Haakon, Hadal, Hankwang, Harishua, Harmi banik111, Harriv, HartzR, Hatesonyericsson, Hauskalainen, Havarhen, Hdt83, Head, Hell9, HelloAnnyong, Hemanshu, Heron, Herr Beethoven, HkCaGu, Hmains, Hooperbloob, Hotwiki, Howardchu, Htanna, Hu12, Hustedcarl, Hydrargyrum, Hylene, IJK Principle, Icewindfiresnow, Icseaturtles, Ilovemymac, Imgaril, Immunmotbluescreen, Imperi, Inspirito, Intelligentsium, Interframe, Inzy, Iohannes Animosus, Iphon, Irishnbears, Irstu, Isfisk, Ithinkicrappedmyself, Ixfd64, J. Sketter, J.delanoy, J36miles, JAAqqO, JCDenton2052, JForget, JGRIFF47, JIP, JLaTondre, JNW, Ja 62, Jackollie, Jak123, Jamcib, Jankratochvil, JayceAndTheNews, Jbsegal, Jclemens, Jdl18, Jeffrey Mall, Jeltz, Jensbn, Jeronimo, Jerry, Jerryseinfeld, Jim, Jimp, JmeSaunders, Jmh, Jni, JoeHinks, Joel7687, John, John Appleseed, JonForst6, Jonathan Hall, Jonik, Jopo, JorgeGG, Joseph Solis in Australia, Jovianeye, Jpk, Jrdioko, Jsysinc, JudasJesus, Julesd, Justin Steele, KC., KFP, KUsam, Kabsingh, Kai Ojima, Kalivd, Kallemax, Kalpesh.v.mistry, Kangaroopower, Katous1978, Kcandrsn, Keegan, Keonne, Ketchup, Khanri01, Kickus, Kirosana, Kjetilho, Kjramesh, Kkm010, Klilidiplomus, KoastalBenefitPromo, Konakalla anurag, Konrad Foerstner, Kosyoboy, Krellis, Kukulcan 7560, Kuponadam, Kwamikagami, LG4761, LLentil, Lafraia, Lamat, Larry laptop, LarryGilbert, Ld100, Leandrod, LeaveSleaves, Lectonar, Leo-loErlahell, Lepensky, Levineps, Lewys, Lfh, Lhotkami, Liamgilmartin, Lightmouse, LilHelpa, LittleDan, Loigenth, Lollomama, Look2See1, Loren.wilton, Lotje, Lovesameer9812, Lowflyingowl, Luigiacruz, Luisvmejia, Lukobe, M3lm4tt, MER-C, MMuzammils, Ma.abilash, Mabdul, Magioladitis, Makedonec28, Makeemlighter, Malhonen, Mani1, Manop, Marc Lacoste, Mardus, Marek69, Mark, Mark Arsten, Marmzok, Martarius, Martin.uecker, Mass09, Mastermind 147, Matrobriva, Matt povey, Mattbr, Mav, Maxim4o, Mayhaymate, Mayurg, McSly, McWika, Megahmad, Menphrad, Mert2000, MetroStar, Mhkay, Mic, Michaelbarr123, Microtony, Mike Rosoft, Mimsie, Minimac, Miraceti, Mirmo!, Misspsyb2, MithrandirAgain, Mkidson, Mmsteelers, Mojei, MominS, Momirt, MonaNL, Monkeynoze, Mowsbury, Mr Stephen, Mr. Met 13, MrFawwaz, MrOllie, MrZoolook, MrsAmethystWay, Mumble45, Murph146, Murphy418, Mysdaao, NKDurrani, Nagytibi, Nakon, Nantasatria, Narcisso, NawlinWiki, NeilN, Neilgravir, Neon white, Newone, Newtechpulse, Nielswik, Nikopolis1912, Ninja5624, Ninjustic, NokiaF, Nokiatrader, Noraft, Norden83, Nscheffey, Nubbly, Nubiatech, Number29, Nuno Tavares, Nuttycoconut, Nwpl, Obli, Ohnoitsjamie, OkakiMCMLXX, Oki putera, Oleksandr Kononenko, Oli Filth, Olivier, Omernos, One, Onlynokia, Onorem, Oo64eva, Ooza, Opraco, Ostralek, Otto2011, Oxymoron83, P0ppe, PTSE, Pablo-flores, Palefire, Parthrana, Pascal.Tesson, Pavel Vozenilek, PeeJay2K3, Pengo, Perohanych, Persia2, Pessi, PeteS, Peter Chastain, Petri Krohn, Pgan002, Pgk, Philip Trueman, Phonefinder2007, Pink Bull, Pista235, Pkadam, PkerUNO, Ploca12, Plokijnu, Pmggp, Pne, Pol098, Pony1142, Ppntori, Prakash.Akshat87, Prari, ProhibitOnions, Prokopov, Prolog, Proudfoot 001, Pseudomonas, Pterre, Puckly, Pudeo, Puneeeetjain, Pupster21, Quackdave, Quattrope, QuiteUnusual, Qwyrxian, Qxz, R'n'B, RTG, RVJ, Radagast83, RadicalBender, Ragityman, Rangoon11, Ratinator, Ratul655, Ravi vadgama, Ravo492, Rcawsey, RedWolf, Reliancepowercoin, Rent A Troop, Retired user 0001, Rettetast, Reuvengrish, RexNL, Rhobite, Richard Harvey, Rickterazor, Riklear, Rjwilmsi, Rklawton, Roamataa, Rob Lindsey, Rob1974, Ron Ritzman, Ronnotel, Rrjanbiah, Rune X2, Ruronimomo, Ruslan0202, Rwalker, S h i v a (Visnu), S3000, SMC, SMP, ST47, Sadunlove, Sai2020, Saimhe, Salilm, Salome5764, Sam Hocevar, Samjuise, Samsara, Sander Säde, Sanixia, Sanjiv swarup, Satchmo2010, Savh, Savvo, Scootey, ScottSteiner, Scupplefish, Seanmilloy, Sebastian Shaw 449, Secaundis, Sedathut, Selimbey, Seqsea, Sfan00 IMG, Sfmammamia, Shadowjams, Sharcho, Shashankbhat, Shaun680, Shawnnicholsonca, Shd, Sherool, Shikker, Shizane, Siafu, SidP, Silpol, Silvah 87, SimonCrowley, SimonThird, Simone, Sinigagl, Skittle, Skyezx, Slavon37, Slo-mo, SmartFace44, SnappingTurtle, Snowolf, Sokratees9, Someguy1221, Soupyjnr, Spangineer, SpigotMap, Squash Racket, Squirtypants, Starblind, Steel, Stephan Leeds, Stephenb, SteveDay, SteveSims, Studioghiblitotoro, Suhailpeerbhai, Suomi Finland 2009, SuperHamster, Sven Manguard, Svgalbertian, Swctg, Syniq, T.O. Rainy Day, THEunique, TYelliot, Tagishsimon, TarzanASG, Tascha96, Tbhotch, Tda, Tedats, Teksosyete, Tellarin, Templetongore, Teqhed, Term061, Test2010, Thaliadrogna, The Man in Question, The Person Who Is Strange, The Random Editor, The Rogue Penguin, The Thing That Should Not Be, TheBlueKnight, TheGreenFaerae, TheYmode, Thebluebeast, Themfromspace, Thingg, Thorvy, Thule, Thunderbrand, TicketMan, Tide rolls, TigerK 69, Timo Honkasalo, Timonoko, Timster69, Tkynerd, Toehead2001, TomB123, Tomisti, Tompagenet, Tomwalden, Tonius, Tony1, Topperfalkon, Tpbradbury, Trakesht, Treschupetes, Tri400, Triage, Trisreed, Tsungik, Tuliopa, Turnstep, Twid, Ufinne, Ulric1313, Ultraviolet scissor flame, Ulysses, Uncle Dick, Unclealex, UpBeat, UpstateNYer, Valentine McKee, Varundbest10, Velella, Venus 9274, Versageek, Vespristiano, Vianello, Vina, Vinayakgole, Vinaywin7, Vininche, Vitund, Vkem, Volgar, Vrenator, Vuo, WIMYV, WJetChao, Wahgujarat, Waycool27, Weatherwax, Welovedoves, Weyes, Wfaulk, Wibbble, Wiki Wonda, Wikipelli, Wikizard2010, Wikizeta, William M. Connolley, WilliamKF, Wimt, WiseOne76, XLerate, Xhienne, Xyzahirm, Yakudza, Yamaaan, Yamamoto Ichiro, Yandman, Yanksox, YellowMonkey, Yousaf465, Yuhani, Yvresgyros, Z10x, ZZninepluralZalpha, Zackwee, Zeno Gantner, Zidonuke, ZirconiumTwice, Zodiak 887, Zpetro, Zsinj, Zunils, امران مجد, محمد المحارب, 55דוד, ஜெலின்கி னேனாந்து, □ □ □ □ , □ □ □ , 1421 anonymous edits

QWERTY *Source*: http://en.wikipedia.org/w/index.php?title=QWERTY *Contributors*: *drew, 098760sam, 12345678998765432l0Z, 216.7.146.xxx, 2D, 321qwerty123, 4Petesake, 5 albert square, 56uywm, 6x7, 94pjg, A. di M., A.h. king, A8UDI, ABarnes94, ADGTHFan, AGK, ARUNKUMAR P.R, AT343, AaronRosenberg, AaronW, Abenyosef, Abomasnow, Access Denied, Acebennett, Addy c86, Adovsdfnfgaerg, Adthebad12, Advancewars177, Aemarques, After Midnight, Agent Smith (The Matrix), Agent007bond, AgentPeppermint, Ahoerstemeier,

Airsoftpwns8500, Ajmas, Alai, Alansohn, Alegoo92, Alexius08, AlexiusHoratius, Alexrl911, Ali, Alisterkhan, Allisonirwin, AlmostReadytoFly, Alphaqt, Alphax, Alyssajay, Amatulic, Ancos, Ancrene wisse, Andre Engels, Andres, Andrewrp, Andrewski, Andy, Andy Dingley, Angmering, Anish 1497, Ann Stouter, Anna Lincoln, Anonymous Dissident, Anschelsc, Antandrus, Anthonybouzi, Antiuser, Arakunem, Arathun, Arkold Thos, Army1987, Arthena, Arthur2045, Aruton, Arzachel, Asdfggrecvbnhgfdxcvb, Asgard337, AthiestsAreCool, Atif.t2, Audriusa, Autiger, Axel*, Azylbac, BOARshevik, B52K, Dan this xd, Danaticus, Dando26, Danhart, Bard of Cornwall, Barefootguru, BarretB, Barrylb, Bart133, Bassbonerocks, Bassplr19, Beesforan, Ben Ben, Bender235, Bento00, Benwildeboer, Bevo, Bige1977, Bill Thayer, BillC, Bjdehut, Blanchardb, BlastOButter42, Blogman222, Bo Lindbergh, Bobblewik, Bobrayner, Boemanneke, Bongwarrior, Bradd, Brammers, Branddobbe, Brianbrady25, Brianski, Brion VIBBER, Bryan Seecrets, Bsroiaadn, Bugs.duggan, Burntsauce, Bushcarrot, Buttheadthedeleter, BytEfLUSh, C.Fred, CNJECulver, CTZMSC3, Calabe1992, Caltas, Can't sleep, clown will eat me, CanadianLinuxUser, Capmango, Capricorn42, CardinalDan, CarmelitaCharm, Carnildo, Carpenoctem, Caster23, Catgut, Cdmarcus, Celi0r, Ceridan, Charles Matthews, Chato, Chill Pill Bill, Chirag, Chrislk02, Christian List, Christian75, Christopher Parham, Chzz, Cinaclov, Circeus, Cleared as filed, Cleduc, Clorox, Closedmouth, Coachs, Codyfitz8, ColdFeet, Colemak fan, Commit charge, Computer97, Conversion script, Coplan, Courcelles, Cpl Syx, Cquan, Creidieki, Crenner, Cretog8, Crispmuncher, Cromarty Maxwell, Crunchy Frog, Cryptic, Cshay, Cst17, Curps, Cyberguy78, DARTH SIDIOUS 2, DFS454, DGaw, DJ Clayworth, DNewhall, DVdm, Dagonweb, Dalemurph, Damian Yerrick, Damicatz, Damonkeyman889944, Dan Polansky, Dan m90, DanMS, DancingPenguin, Danelo, Danny247, Danski14, DanteHadoken, Darkwind, Darrel francis, Darthnater13, DarwinPeacock, Dave souza, Davemcarlson, DavesPlanet, David Gale, David Gerard, David Newton, DavidRF, Davidhorman, Davwi379, Dawn Bard, Dawnseeker2000, Dcandeto, Dd42, Deadfeet, Deathsythe, Deflective, Delval3269, Demonic Locust, Denelson83, Denisutku, DennyColt, Depaderico, DerHexer, Derekbd, Derex, Dffgd, Diannaa, Dichdiger, Diggernet, Digitaleon, Dinoguy1000, Directorschair, Discospinster, Dismas, DivineBurner, Dodo bird, Dodo von den Bergen, DoubleBlue, DrZeus, Dreftymac, Dshatg9uieasrg8our, Dskluz, Duffman, Dysprosia, Dziban303, Ecjmartin, EdH, Edgar181, EdgeOfEpsilon, Edward, Eebster the Great, Eeekster, Eequor, Ektoric, ElBenevolente, Elassint, ElationAviation, Elcobbola, Elium2, Endlesspastaolivegarden, Endlessshrimpredlobster, Enviroboy, Epbr123, Epson291, Epugachev, Ericphillips, Erik9, ErkinBatu, Escape Orbit, Eskovan, Esperant, Espoo, Essjay, Etineskid, Eurleif, Everyone Dies In the End, Evil saltine, Ewx, Ex nihil, Excirial, Exidor, Explicit, FAKEFAKEFAKEFAKEFAKEFAKEFAKEFAKE, Falcon8765, Falcon9x5, Fantom, Faradayplank, Favonian, Fayenatic london, Fear the hobbit, Felix Wiemann, Fergie4000, Fetchcomms, Fieldday-sunday, Filelakeshoe, Finalius, Finlay McWalter, Firewall this, Fischer.sebastian, Fischguy111, FisherQueen, FleetCommand, Flip619, Flipping Mackerel, Flokarti, Floobman, Fluffernutter, Flyingsquirrel, Fowlerism, Fran McCrory, FrancoGG, Frankie.yuen, Frazzydee, Freakofnurture, Frood, Fuhghettaboutit, Fuiq3, Furrykef, Futanari, Futurebird, Fæ, GBug2, Gadfium, Gaganspidey, Gaia Octavia Agrippa, Gamer007, Gamerdude94, Gargantua cat, Gazimoff, Geke, Geneb1955, GeorgeLouis, Gerbrant, Get It, Gilliam, Ginkgo100, Gjgfuj, Gjking, Glacialfox, Gladlyplaid, Glane23, Glen, Glennwells, God of War, Gogo Dodo, Golbez, GoldKanga, Gomm, GoneAwayNowAndRetired, Goodmanj, GorillaWarfare, Gorkaazk, Gosub, Gpia7r, Gracenotes, Grafen, Graham87, GranterOfMercy, Greendickinson, Greggsparks, Gregory Heffley, Gregozzy, Grim23, Grizzly37, Gwib, Gökhan, H005, HJ Mitchell, HalfShadow, Hall Monitor, Hallo1234, Hamtechperson, Happynoodleboycey, Harry, Hawaiiboy99, Hayesmen, HazardTheory, Hazardous Matt, Hdt83, Hetenyid, HexaChord, Hippietrail, Hirudo, Hodul, Homelessman123123123, Hongooi, Hoosteen5, Hoovernj, Howcheng, Husond, Huw Powell, Hydroxicacid, I am a freakazoid, I am me93, Iameukarya, Idh0854, Idleguy, Immunize, Indiana State, Indon, Inhumandecency, Interlingua, IntrigueBlue, Iris4, IronGargoyle, Ispy1981, It Is Me Here, Ivan, Ixfd64, Izwalito, J.delanoy, J10696, JCarriker, JDDJS, JDP90, JFreeman, JHP, JJLeahy, Jackfork, JadeBoco, Jake1192, Jaktv, Jamesoe, Jammycakes, Janeway216, Jaranda, Jayc52925, Jayjg, Jcw69, Jecar, Jeddac, JediLofty, Jennieo, JeremyMcCracken, JerryFriedman, Jessylou, Jfm48, Jgterzian97, Jh51681, Jiawei18, JidGom, Jim Douglas, Jim.henderson, Jim1138, Jimmy da tuna, Jlguenego, Jmwilkinson, Jncraton, Joe1337, John, John254, JohnCD, Jojhutton, Jonathan de Boyne Pollard, Jonesykat, Joseph Solis in Australia, Josephina24, JoshDuffMan, Jossi, Joyous!, Jpp42, Js4362, Ichadias, Itedor, JuJube, Juliancolton, Jusdafax, Justin"drool"bieber, Jwink3101, KNHaw, Ka Faraq Gatri, Kaiba, Karen nicole, Katalaveno, Katieh5584, Kazrak, Kbdank71, Kbrose, Keakealani, Kevinmon, KeyStorm, Khairul hazim, Khuday, Kilgara, Killiondude, Kilom691, King Lopez, King of Hearts, Kingerties, Kingpin13, Kubigula, Kudret abi, Kukini, Kuru, Kwamikagami, Kwpolska, Kyle1278, Kylemcinnes, L Kensington, LarRan, Larry V, Lateg, Lbr123, Lcgarcia, LeaveSleaves, Lee Daniel Crocker, Leithp, Lemmon Juice, LemonLion, Leo11, Lestrade, Leyo, Lightmouse, Lights, Linnormlord, LittleDan, Livajo, LjL, Ljohnson2928, Lonelydarksky, Loodog, LoonyLeif, Love Aaron, Lovinoven, LtPowers, Luk, Luluroks, Lurlock, M.thoriyan, MCBastos, MER-C, MK8, MZaplotnik, MacMed, Maceis, Madhero88, Maebmij, Magicmarc, Magister Mathematicae, Magnus.de, Magore, Mailer diablo, Majora4, Makusu2, Malo, Manlymanthe3rd, Marek69, Mariah8787, Mario.brown112, MarkSutton, Markreidyhp, MartinUK, MashedTaters, Matrix8110, Matt.smart, Mattgirling, Matthew Yeager, Matthewcieplak, Mattman9009, MauchoEagle, Maximaximax, Maxis ftw, Maxí, Mbarbier, Mbecker, McSly, Mcintyem, Meand, Med, Melongrower, Melsaran, Menchi, Mentifisto, Mephistophelian, Message From Xenu, Metalmario128, Mets501, Mgenerous, MichaelMaggs, Microtony, Mightyeldude, Mike 7, Mikejr0615, Millionsandbillions, Minimac, Minna Sora no Shita, Miquonranger03, Miremare, Mm40, Mnop64, Money4nothing, Monkbel, Monterey Bay, Monty INIT, Mowinx, Mr Stephen, Mr Twix, Mr. Stradivarius, Mr.GL399, MrZoolook, Mrand, Mrg3105, Mseliw, Mspraveen, Muglug, Muslim Editor, Mussavcom, Myanw, Mygerardromance, Myscrnnm, Mzajac, N96, NERD 3.14159, NHRHS2010, NPalmius, Natwebb, NawlinWiki, Nbarth, Nemzai, NerdyScienceDude, NetRolller 3D, Netizen, NewEnglandYankee, NickelShoe, Nickg, Nickj, Nicmo9462, Nietzscheanlie, Nifky?, NightFalcon90909, NikiAnna, Nikola Smolenski, Nils peterson, Nlu, Nonubb, NotMuchToSay, Nothingofwater, Novulus, NuclearWarfare, Nudecline, Nufy8, Nunh-huh, Nutiketaiel, Octan, Ojigiri, Oldlaptop321, Olivier, Olleicua, Omicronpersei8, Omphaloscope, Onebravemonkey, Opelio, Orange Suede Sofa, Orca1 9904, Ost316, Ottawa4ever, Overtwitch2, Oxymoron83, Polyglut, P199, PHenry, Pirish, Panel Guy, Paul Foxworthy, Paul the alien, Pawyilee, Pdcook, PedroFonini, Per Hedetun, Perey, Person23qwerty45, Peter S., Petiatil, Petra123, Pfranson, Pgk, Pharaoh of the Wizards, PhilHibbs, PhilKnight, Philip Trueman, Phils, Phoenix2, Piano non troppo, Pill, Pinkadelica, Pinkbeauty813, Plujfaf, Pmsyyz, Pn57, Pol098, Polkasucks, Pparazorback, Praefectorian, Protolink, Proword, PseudoSudo, Pwizard02, QUINTIX, QWERTYuiop1234, Qero, Qet135, Quadpus, QuadrivialMind, Quintillus, Qutezuce, Qwerty355, Qwertyc, Qwertymang, Qwertyuiop11111, Qxz, RHBridges, RJM, Radiant chains, Rafpilotdavid, Raiderfan038, RandomStuff, Randomguy99, Randy6767, Rangi42, Rani Lueder, Raoul2, Raslafor, Rattermanjason, Ravedave, Raycy, Readgirl, Recognizance, Red Director, Redvers, Reedy, Relayer250, Reliableforever, Remy B, Rennocbitgod, Rettetast, RexNL, Rice123, Rich Uncle Skeleton, Richard D. LeCour, Rickyj112, Rjd0060, Rjwilmsi, RkOrton, Rklawton, Rmhermen, Robert Brockway, RobertG, Robesquie87, Robust Physique, RockRichard, Ronbo76, Ronnjones, RonkU1, Ronz, Rossami, Rottyedit016, Route 183, RoyBoy, Royboycrashfan, Rrburke, Rtyq2, Rubo77, Rudjek, RufusThorne, Rum runner90, RussBlau, S-Ranger, SFC9394, SFGiants, SKREAM, SMP, SOS48, SYSS Mouse, Sade, Salmanazar, Salt Yeung, Samm2112, Samrees2007, Samtheboy, Samuelsen, Sango123, Sanmartin, Sarefo, Saros136, Sarranduin, Satyashree, Saxophobia, Scarian, Sceptre, SchnitzelMannGreek, SchuminWeb, Scientistpatrick, Scientizzle, ScottyBerg, SeanMack, Seaphoto, SebRovera, SeldomSerious, Shabs09, Shakey-Lo, Shanel, Sharcho, Shawn in Montreal, Shifty(clue), Shifty401, Shirt58, Shreevatsa, Shrish, Shubinator, Siebrand, Sietse Snel, Silivrenion, Simetrical, SimonP, Sir Nicholas de Mimsy-Porpington, SirGrant, Siroxo, Sjb90, Skarebo, Skootles, SkyLined, Slakr, Slappy381, Slawojarek, Smalljim, Smegpt86, Smileyface11945, Sn0wflake, Snakeyes (usurped), Snalwibma, Snaxe920, Snthdiueoa, Soccerbuster016, Soccerdude999, Soliloquial, Some jerk on the Internet, Someguy1221, Spellcast, SpuriousQ, Ssr, Ssteedman, Star9837, Steamboy, SteinbDJ, Stephan Leeds, Stephenb, Stevage, Steve3849, Stickyfox, Stratadrake, Stratocracy, Stroppolo, Stubblyhead, Studerby, Stupid Corn, Subrih, SunDragon34, Super Mario, Super-Magician, Superbeecat, SupermanReturns, Suriel1981, Survival705, Svanslyck, Sverre avnskog, Svick, SwisterTwister, Swpb, Synchronism, T-borg, THB, Tagishsimon, Tanaats, Tarquin, Tavilis, Tbhotch, Techman224, Technopat, Tempodivalse, Temptinglip, Tenofour, Tetraedycal, ThaddeusB, Thaurisil, The Anome, The Chinchou, The Epopt, The Haunted Angel, The Maid, The Man in Question, The Sanest Mad Hatter, The Thing That Should Not Be, The foot clan, The king 556, The silent gnome, The wub, TheDJ, TheProject, TheProphetTiresias, TheRedPenOfDoom, Thecoolguyman123, Theda, ThefirstM, Thegayguy1337, Theoutstrip, Thesis4Eva, Thewayforward, Thingg, Thumperward, Tide rolls, TigerK 69, TimMagic, Tiptoety, Titanium Dragon, Tjmayerinsf, Tkgd2007, Tobias Bergemann, TobyHung1234, Tohd8BohaithuGh1, Tom harrison, Tom616, Tomasheasly, Tombomp, Tomekemon, Tommy2010, Tony Hunter, Trashday03, Travv0, Tregoweth, TresRoque, Tresiden, True Pagan Warrior, Turanyuksel, Turtleboy, Twang, Twilly41, Typenolies, Tysto, UberScienceNerd, Ucanlookitup, Ukexpat, Umidrb, Ummit, Uncle Dick, Uncle Milty, Unknownwarrior33, Urbane Legend, Urhixidur, Uruiamme, Usability1, Utcursch, UtherSRG, Utsutsu, Valentinian, Valley2city, Vanished user 5zariu3jisj0j4irj, Velella, Versus22, Vinhtantran, Vinitp2, Volt4ire, Vquex, VsevolodKrolikov, Wa3frp, Waggers, Walter Görlitz, Wangi, Wapp, Wavelength, Wayward, Wernher, Wesley, Weslinboy2010, Wesman1995, West.andrew.g, Whisky drinker, Why did you do it, Wiff&Hoos, Wiikipedian, WikHead, Wiki alf, Wiki13, WikiBum, Wikieditor0410, Wikipelli, Wikiwow, Will Beback Auto, Will Smith Bro, William Avery, Williamsj2, Willking1979, Wilsonang123, Wiooiw, Wknight94, Woohookitty, Wordbuilder, Wtmitchell, Wyatt915, Xezbeth, Xploita, Yamamoto Ichiro, Yes0song, Yf metro, Yoda Drinks Soda, Yomangani, Yourloveisvain, Yugsdrawkcabeht, Zap3510, Zarniwoot, Zavav, ZeroOne, ZooFari, Zorro CX, Zsinj, Ztobor, Zurishaddai, Zxzxzxg, Zzuuzz, Ĝaaw, □□□□□, 1970 anonymous edits

Entry_Level *Source*: http://en.wikipedia.org/w/index.php?title=Entry_Level *Contributors*: Green Tentacle, Joshua Issac, KathrynLybarger, Pegship, Ynhockey, 1 anonymous edits

Secure_Digital *Source*: http://en.wikipedia.org/w/index.php?title=Secure_Digital *Contributors*: 24frames, 81120906713, 99bluefoxx, 9allenride9, A Man In Black, AVRS, Aab298347927384, Adam Schloss, Adam2288, Adsp, Aeons, Agateller, Agrowarrior, Aikenware, Airplaneman, Akaustav, Alansohn, Aldie, AlexJ, Alexbuster500, Alexthekiwi, Allaun, Alphaman, Alvis, Amezcackle, Amoiwangjy, Amram99, Andrevan, Andrew sh, AndrewKepert, Andrewpmk, Andru nl, Angela, Angerdan, Animehawaii, Appraiser, Aqn, Aribronstein, Armando, Arrowcatcher, Asim18, AssetBurned, Atanasov, Atompowered, Austinmurphy, Autoandragogist, Axcelis555, AzaToth, BENNYSOFT, BPM, Bastique, Bdelisle, Beao, Bearcat, Beeline23, Beland, Bergsten, Bgcaus, Bholstege, Billgordon1099, Binba, Bjdehut, Blanchardb, BlindWanderer, Blorg, Bobblewik, Bobianite, Boboangel, Bomazi, Bookandcoffee, Bostwickenator, Brer vole, Brian Patrie, Brian-L, Briandgregory, Brianski, Brouhaha, Buster2058, Buuneko, C 1, CDV, Calcprogrammer1, Caleson, Calor, Can't sleep, clown will eat me, Canrocks, Carreon77, CaseyPenk, Caspertheghost, Catskul, Cdrum, CesarB, Chandu15, CharlotteWebb, Chealer, Cherkash, Chris the speller, ChrisCork, Chrisjj3, Christian75, Ckmac97, Climber22, Clorox, Cmdrjameson, Compellingelegance, Compuguy1088, Coneslayer, Consumed Crustacean, Cootiequits, Copysan, Corge, Coughinink, CrookedAsterisk, Csendesmark, Cyawman, Cybercobra, Cybertai, CyclePat, D0nj03, DMahalko, DaBunny42, Dabomb87, Dah31, Dale Arnett, Damian Yerrick, Dan100, DanMS, Dancingonmice, Dancter, Dandv, Daniel Newby, Danielt998, Dankru, Darin-0, Darkxsun, Darrien, Daveswagon, David from Downunder, David spector, David.Monniaux, Davidswelt, Dawd, Deineka, Dennbruce, DervishD, Dhanashekar, Djm1279, Dlrohrer2003, Doc glasgow, DocWatson42, Doctorno, Dogcow, Drabant, Draconiator, Dreadstar, Drmies, Dustin Howett, Długosz, EJSawyer, Echtner, Ed Poor, Eftpotrm, Electricnet, Electron9, ElementFire, Elemesh, Elvey, Elwilke, EncMstr, Ente75, Erencexor, Ericzhang789, Esowteric, Eurleif, Evice, Excirial, Exearly, ExplicitImplicity, ExportRadical, FCartegnie, Fast healthy fish, Feedmecereal, Feydey, Flarn2006, FlashSheridan, Flasher, Fleminra, Flowerflower, Fmillour, Franciscouzo, Freddyzdead, Fresheneesz, Fudoreaper, Fullerene, Funandtrvl, Furrykef, Fvw, Gadol87, Gaselvin, Ge0rge, Geekosaurus, Gglockner, Ghaib, Glaw10, GodGell, Goosnarrggh, GraemeLeggett, Grafen, Graham87, GregorB, Grunt, Guy Macon, HDCase, Haipa Doragon, Hankwang, Hansa11, Harlekeyn, Harvester, Hebrides, Henry W. Schmitt, Heron, Heybales, Hgrosser, Hoagg, Hooperbloob, HoserHead, Hurk87, IAmAI, IW.HG, IceHunter, Ichimonji10, IkonicDeath, Iliev, Impi, Intelliot, InternetMeme, Irayo, Ismouton, J.delanoy, JJLatWiki, Jachim, JackTinWNY, Jafeluv, Jahoe, James JK, JamesAM, JamesTeterenko, Jdowland, Jeffq, Jerryobject, Jhsounds, Jidanni, Jim McKeeth, Jim.henderson, Jj2235, Jlslspam, Jnavas, Joaoplim, Jocunddus, Joe Sewell, John of Reading, JohnAHind, JohnAlexan, Johnteslade, Jordan Brown, JordoCo, Jory, Josi.ow, JudyJohn, Justin Ormont, Jvr725, Jw21, Kartano, Karthikndr, Kashmiri, Kaszeta, Kazrak, Kbdank71, Kiimmyluff, Kinema, Kingpin13, Kjohna, Klkmiooo, KnowBuddy, Koolman2, Korj.by, Kozuch, KryptoCleric, KsprayDad, Kungfujoe, Kungming2, Kusunose, Kvng, LM1987, Lamrock, Lavenderbunny, Lbecque, Le Déchaîné, LeoO3, Leszek Jańczuk, Liaocyed, Lightmouse, LilHelpa, Lindosland, Little Professor, Littleendian, Lord Nightmare, Lordsatri, LostLeviathan, LovesMacs, Lucasreddinger, MER-C, MMuzammils, Macrakis, MadisSepp, Manuishino, MarXidad, Marchash, Marianocecowski, Mark Renier, Martarius, Matthiaspaul, Maury Markowitz, Maxim Leyenson, Maxí, Mboverload, Megazero, Memorysuppliers, Metallica10, Michael L. Kaufman, MichaelJanich, Microsofkid, Mihai Capotă, Mikus, Misterkillboy, Mju7nhy6, Mmj, Mnts, Modster, Moreati, Mortense, Moxfyre, MrBurns, Mram80, Mrdungx, Mrzaius, Msaunier, Mwarren us,

Myke2020, N1truX, N5iln, NEMT, Nagualdesign, Nahuel.carvajal, Nasukaren, Neelix, Neilka, NerdyNSK, Neurolysis, Newzack, Nil Einne, Nintendude, Nixdorf, Nnan, Noir, Noq, Notmicro, Now3d, Ntsimp, Nulltransfer, Nurg, Nv8200p, OMA2k, Oahiyeel, Oe1kenobi, Offbeatcinema, Ojigiri, Oliverdl, Omeganian, Omegatron, Oxymoron83, Pabouk, Paranoid, PatrikR, Peace81, Peaceninja28, Peter Campbell, Peter S., Peterl, Peterrobbemond, Peyre, Photographerguy, Photoguy439, Phr, Pickmynose1999, PierreOssman, Pinkevin, Piper8, Placi1982, Plasmaroo, Pranjal87, ProhibitOnions, Psharrock, Psiddall, Quaeler, Queefycreatures, Ravedave, Ravenperch, Razor2988, Rcawsey, RealGrouchy, Reaperducer, Rebroad, RichardTector, Riddley, RingtailedFox, Rjc34, Rjquillin, Rjwilmsi, Rlcantwell, Rmsuperstar99, Rmunn, Rofl cawpters, Rohan Jayasekera, RoseTech, Rosenbluh, Rstoplabe14, Ruud Koot, SD Card Version 2.Wiki.0, SF88, SGBailey, SMD915, SPKirsch, SRG275, Sagarsavla, Saimhe, Saipraneethn, Saltmiser, Sam8, SanGatiche, Sanalmg, Sandymac, Saravanants, Sasan.j, Saverworld2, Sbmeirow, SchnitzelMannGreek, Scientus, Scotty jasper2000, SeanAhern, Seaphoto, Sebadee, Sedimin, Sergei, Shadowjams, Shanes, SidP, Sin-man, Skathol, Skierpage, Skyykj, Slakr, Slashme, Slicing, Smack, Smileyborg, Smyth, Snolygoster, Speedevil, Spike-from-NH, Spikesagal, Stanleyivan, Starnestommy, Steinsomers, Stephan Leeds, StephenFerg, Stevage, Stuston, Sunny house, Surf243, Suriel1981, Suruena, Szzuk, TAC-3, TCav, Tacvek, Tamiledu, Tarquin, TedPavlic, The Giant Puffin, The Thing That Should Not Be, TheDoober, Themoment, ThevillagesmithE, Thewikipedian, Thunderbird2, Thunderpenguin, TimSE, Timharwoodx, Tintenfischlein, Tiredofscams, Tkgd2007, Tmansour, TobleRone, Todd Vierling, Toehead2001, Tomgibbons, Tone, Tony1, Tonyhawz, Totakeke423, Towel401, Tpbradbury, Triscal1990, Tristan Schmelcher, Tweenk, Twinmostech, U4ia74, UKER, UU, UberMan5000, UltraMagnus, Vdcappel, Versus22, Vexorg, Viggio, Viol8or, Vladsinger, VoxLuna, Waffle, Wammes Waggel, Wanderer099, Wbt92, Wbuch, Weezee, Wernher, WhiteDragon, Whitis, Wibbble, WikHead, WinTakeAll, Wine Guy, Wirbelwind, Wjejskenewr, Wjw0111, Wolfling, WriterHound, Wrlee, Wtshymanski, Wuhwuzdat, XP1, Xapplimatic, Xdddex, Xenon54, Xmhd, Xonicx, Xpclient, Yaniv Kunda, Yealout, Z hosen, Zarcillo, Zarenor, Zbrahead91, Zerpent, Zodon, Zoicon5, Zsexdrcft, Zurotai, Zviangi, Érico Júnior Wouters, Žiga, 1185 anonymous edits

Image Sources, Licenses and Contributors

File:Nokia x2-01.jpg *Source*: http://en.wikipedia.org/w/index.php?title=File:Nokia_x2-01.jpg *License*: Creative Commons Attribution-Sharealike 3.0 *Contributors*: User:Pakshya

File:Nokia 6020.jpg *Source*: http://en.wikipedia.org/w/index.php?title=File:Nokia_6020.jpg *License*: Creative Commons Attribution 2.0 *Contributors*: rodrigo senna

Image:Nokia 6300.jpg *Source*: http://en.wikipedia.org/w/index.php?title=File:Nokia_6300.jpg *License*: Public Domain *Contributors*: Original uploader was Racklever at en.wikipedia

File:Symbian S40 v10.80.png *Source*: http://en.wikipedia.org/w/index.php?title=File:Symbian_S40_v10.80.png *License*: Creative Commons Attribution-Sharealike 3.0 *Contributors*: Blackvienna, 1 anonymous edits

File:Nokia_N9.jpg *Source*: http://en.wikipedia.org/w/index.php?title=File:Nokia_N9.jpg *License*: Creative Commons Attribution-Sharealike 3.0 *Contributors*: User:Animist

File:World_Wide_Smartphone_Sales.png *Source*: http://en.wikipedia.org/w/index.php?title=File:World_Wide_Smartphone_Sales.png *License*: GNU Free Documentation License *Contributors*: Smartmo

File:World_Wide_Smartphone_Sales_Share.png *Source*: http://en.wikipedia.org/w/index.php?title=File:World_Wide_Smartphone_Sales_Share.png *License*: GNU Free Documentation License *Contributors*: Smartmo

File:World_Wide_Smartphone_Startup_Share.png *Source*: http://en.wikipedia.org/w/index.php?title=File:World_Wide_Smartphone_Startup_Share.png *License*: GNU Free Documentation License *Contributors*: Smartmo

File:World_Wide_Smartphone_Market_Share_outlook.png *Source*: http://en.wikipedia.org/w/index.php?title=File:World_Wide_Smartphone_Market_Share_outlook.png *License*: GNU Free Documentation License *Contributors*: Smartmo

File:Galaxy Nexus smartphone.jpg *Source*: http://en.wikipedia.org/w/index.php?title=File:Galaxy_Nexus_smartphone.jpg *License*: Creative Commons Attribution 2.5 *Contributors*: Faramarz, MB-one, Nezdek, SF007, 2 anonymous edits

File:IPad-WiFi-1stGen.jpg *Source*: http://en.wikipedia.org/w/index.php?title=File:IPad-WiFi-1stGen.jpg *License*: Public Domain *Contributors*: Evan-Amos

File:Nokia N8 Smartphone.jpg *Source*: http://en.wikipedia.org/w/index.php?title=File:Nokia_N8_Smartphone.jpg *License*: Public Domain *Contributors*: Binoyjsdk, Editor182, Gauravjuvekar

File:Nokia wordmark.svg *Source*: http://en.wikipedia.org/w/index.php?title=File:Nokia_wordmark.svg *License*: Trademarked *Contributors*: Apalsola, Bencmq, ELeschev, Editor182, Hautala, Yarl, 1 anonymous edits

File:Decrease2.svg *Source*: http://en.wikipedia.org/w/index.php?title=File:Decrease2.svg *License*: Public Domain *Contributors*: Sarang

Image:Fredrik Idestam.png *Source*: http://en.wikipedia.org/w/index.php?title=File:Fredrik_Idestam.png *License*: Public Domain *Contributors*: -Majestic-, Apalsola, Jpk, Martin H.

Image:Leo Mechelin (cropped).png *Source*: http://en.wikipedia.org/w/index.php?title=File:Leo_Mechelin_(cropped).png *License*: Public Domain *Contributors*: -Majestic-, Martin H.

File:Nokia 150 and nokia 1100.jpg *Source*: http://en.wikipedia.org/w/index.php?title=File:Nokia_150_and_nokia_1100.jpg *License*: Public Domain *Contributors*: Lvova Anastasiya (Львова Анастасия, Lvova)

File:Nokia booklet 3g-10 (3949263497).jpg *Source*: http://en.wikipedia.org/w/index.php?title=File:Nokia_booklet_3g-10_(3949263497).jpg *License*: Creative Commons Attribution-Sharealike 3.0 *Contributors*: http://thenokiablog.com/Reposted by Mark Guim from United States

File:Nokia HQ.jpg *Source*: http://en.wikipedia.org/w/index.php?title=File:Nokia_HQ.jpg *License*: Creative Commons Attribution-Sharealike 3.0,2.5,2.0,1.0 *Contributors*: -Majestic-

File:Nokia evolucion tamaño.jpg *Source*: http://en.wikipedia.org/w/index.php?title=File:Nokia_evolucion_tamaño.jpg *License*: Public Domain *Contributors*: Jorge Barrios

File:All 9xxx.png *Source*: http://en.wikipedia.org/w/index.php?title=File:All_9xxx.png *License*: GNU Free Documentation License *Contributors*: derivative work: -Majestic- (talk) All9xxx.jpg: Original uploader was R@y at de.wikipedia (2004-04-03)

File:Nokia N8 (front view).jpg *Source*: http://en.wikipedia.org/w/index.php?title=File:Nokia_N8_(front_view).jpg *License*: Public Domain *Contributors*: Editor182, Vivinnl, X-Pilot

File:SymbianWMWP7USMarketShare.png *Source*: http://en.wikipedia.org/w/index.php?title=File:SymbianWMWP7USMarketShare.png *License*: Creative Commons Attribution-Sharealike 3.0 *Contributors*: User:Enemenemu

File:Nokia E55 01.jpg *Source*: http://en.wikipedia.org/w/index.php?title=File:Nokia_E55_01.jpg *License*: Creative Commons Attribution-Sharealike 2.0 *Contributors*: James Nash

File:Nokia N900-1.jpg *Source*: http://en.wikipedia.org/w/index.php?title=File:Nokia_N900-1.jpg *License*: Creative Commons Attribution-Sharealike 3.0,2.5,2.0,1.0 *Contributors*: User:Ilya Voyager

File:Nokia E90 communicator.JPG *Source*: http://en.wikipedia.org/w/index.php?title=File:Nokia_E90_communicator.JPG *License*: Creative Commons Attribution 3.0 *Contributors*: Georgy90

File:Nokia5800xpress.png *Source*: http://en.wikipedia.org/w/index.php?title=File:Nokia5800xpress.png *License*: Creative Commons Attribution-Sharealike 3.0 *Contributors*: Nokia_5800_XpressMusic_Browser.jpg: Jupter-manzana derivative work: TheAdam0s (talk)

File:Flag of Canada.svg *Source*: http://en.wikipedia.org/w/index.php?title=File:Flag_of_Canada.svg *License*: Public Domain *Contributors*: Anomie

File:Flag of Finland.svg *Source*: http://en.wikipedia.org/w/index.php?title=File:Flag_of_Finland.svg *License*: Public Domain *Contributors*: Drawn by User:SKopp

File:Flag of the United States.svg *Source*: http://en.wikipedia.org/w/index.php?title=File:Flag_of_the_United_States.svg *License*: Public Domain *Contributors*: Anomie

File:Flag of Switzerland.svg *Source*: http://en.wikipedia.org/w/index.php?title=File:Flag_of_Switzerland.svg *License*: Public Domain *Contributors*: User:Marc Mongenet Credits: User:-xfi- User:Zscout370

File:Flag of Germany.svg *Source*: http://en.wikipedia.org/w/index.php?title=File:Flag_of_Germany.svg *License*: Public Domain *Contributors*: Anomie

File:Flag of Norway.svg *Source*: http://en.wikipedia.org/w/index.php?title=File:Flag_of_Norway.svg *License*: Public Domain *Contributors*: Dbenbenn

File:Flag of France.svg *Source*: http://en.wikipedia.org/w/index.php?title=File:Flag_of_France.svg *License*: Public Domain *Contributors*: Anomie

File:Nokia Connecting People.svg *Source*: http://en.wikipedia.org/w/index.php?title=File:Nokia_Connecting_People.svg *License*: Public Domain *Contributors*: Original uploader was -Majestic- at en.wikipedia

File:Nokian logo.svg *Source*: http://en.wikipedia.org/w/index.php?title=File:Nokian_logo.svg *License*: Public Domain *Contributors*: Z10x at en.wikipedia

File:Nokian pääkonttori Keilaniemessä.jpg *Source*: http://en.wikipedia.org/w/index.php?title=File:Nokian_pääkonttori_Keilaniemessä.jpg *License*: GNU Free Documentation License *Contributors*: J-P Kärnä

Image:QWERTY keyboard.jpg *Source*: http://en.wikipedia.org/w/index.php?title=File:QWERTY_keyboard.jpg *License*: Creative Commons Attribution-Sharealike 3.0 *Contributors*: MichaelMaggs

File:Loudspeaker.svg *Source*: http://en.wikipedia.org/w/index.php?title=File:Loudspeaker.svg *License*: Public Domain *Contributors*: Bayo, Gmaxwell, Husky, Iamunknown, Mirithing, Myself488, Nethac DIU, Omegatron, Rocket000, The Evil IP address, Wouterhagens, 20 anonymous edits

File:Continental Standard typewriter keyboard - key detail.jpg *Source*: http://en.wikipedia.org/w/index.php?title=File:Continental_Standard_typewriter_keyboard_-_key_detail.jpg *License*: Creative Commons Attribution-Sharealike 3.0 *Contributors*: Continental_Standard_typewriter_keyboard.jpg: Sommeregger derivative work: Nils von Barth (nbarth) (talk)

Image:QWERTY 1878.png *Source*: http://en.wikipedia.org/w/index.php?title=File:QWERTY_1878.png *License*: Public Domain *Contributors*: C.L. Sholes

Image:Qwerty.svg *Source*: http://en.wikipedia.org/w/index.php?title=File:Qwerty.svg *License*: Creative Commons Attribution-ShareAlike 3.0 Unported *Contributors*: Azaghal of Belegost, Davepape, Doggitydogs, EugeneZelenko, Gennaro Prota, GeoGab, HenkvD, LjL, Moberg, Mozaika2, Mysid, Pyerre, Simo Kaupinmäki, TZM, Ymulleneers, Überraschungsbilder, 22 anonymous edits

Image:Nokia E55 01.jpg *Source*: http://en.wikipedia.org/w/index.php?title=File:Nokia_E55_01.jpg *License*: Creative Commons Attribution-Sharealike 2.0 *Contributors*: James Nash

File:SD Cards.svg *Source*: http://en.wikipedia.org/w/index.php?title=File:SD_Cards.svg *License*: Trademarked *Contributors*: Tkgd2007, □ □

File:MicroSD MemoryCard 002.jpg *Source*: http://en.wikipedia.org/w/index.php?title=File:MicroSD_MemoryCard_002.jpg *License*: Creative Commons Attribution-ShareAlike 3.0 Unported *Contributors*: Kropsoq

File:TwinMOS 16gb sdhc.jpg *Source*: http://en.wikipedia.org/w/index.php?title=File:TwinMOS_16gb_sdhc.jpg *License*: Creative Commons Attribution-Sharealike 3.0 *Contributors*: Z hosen

File:SDXC.svg *Source*: http://en.wikipedia.org/w/index.php?title=File:SDXC.svg *License*: Public Domain *Contributors*: SD Card Association

File:HP PhotoSmart SDIO Kamera.jpg *Source*: http://en.wikipedia.org/w/index.php?title=File:HP_PhotoSmart_SDIO_Kamera.jpg *License*: Creative Commons Attribution-Sharealike 2.0 *Contributors*: Afrank99

File:USB-SD-Cards.jpg *Source*: http://en.wikipedia.org/w/index.php?title=File:USB-SD-Cards.jpg *License*: Public Domain *Contributors*: AssetBurned

File:Colors of SD card.jpg *Source*: http://en.wikipedia.org/w/index.php?title=File:Colors_of_SD_card.jpg *License*: Creative Commons Zero *Contributors*: User:SreeBot

File:SDHC HD C16 32GB .jpg *Source*: http://en.wikipedia.org/w/index.php?title=File:SDHC_HD_C16_32GB_.jpg *License*: Creative Commons Attribution-Sharealike 3.0 *Contributors*: ACGFAN

File:SDHC Speed Class 2.svg *Source*: http://en.wikipedia.org/w/index.php?title=File:SDHC_Speed_Class_2.svg *License*: Creative Commons Attribution-Sharealike 3.0 *Contributors*: User:Kizar

File:SDHC Speed Class 4.svg *Source*: http://en.wikipedia.org/w/index.php?title=File:SDHC_Speed_Class_4.svg *License*: Creative Commons Attribution-Sharealike 3.0 *Contributors*: User:Kizar

File:SDHC Speed Class 6.svg *Source*: http://en.wikipedia.org/w/index.php?title=File:SDHC_Speed_Class_6.svg *License*: Creative Commons Attribution-Sharealike 3.0 *Contributors*: User:Kizar

File:SDHC Speed Class 10.svg *Source*: http://en.wikipedia.org/w/index.php?title=File:SDHC_Speed_Class_10.svg *License*: Creative Commons Attribution-Sharealike 3.0 *Contributors*: User:Kizar

File:Sdadaptersandcards.jpg *Source*: http://en.wikipedia.org/w/index.php?title=File:Sdadaptersandcards.jpg *License*: Creative Commons Attribution 3.0 *Contributors*: Original uploader was Atanasov at en.wikipedia Later version(s) were uploaded by Brybry26 at en.wikipedia.

File:Canon hf100 with memory card.jpg *Source*: http://en.wikipedia.org/w/index.php?title=File:Canon_hf100_with_memory_card.jpg *License*: Creative Commons Attribution-Sharealike 3.0 *Contributors*: Mikus (talk). Original uploader was Mikus at en.wikipedia

File:Kingston_19-in-1_memory_card_reader.jpg *Source*: http://en.wikipedia.org/w/index.php?title=File:Kingston_19-in-1_memory_card_reader.jpg *License*: Creative Commons Attribution-Sharealike 3.0 *Contributors*: King of Hearts

File:Arduino Ethernet Shield (pre-production sample).jpg *Source*: http://en.wikipedia.org/w/index.php?title=File:Arduino_Ethernet_Shield_(pre-production_sample).jpg *License*: Creative Commons Attribution-Sharealike 2.0 *Contributors*: Matt Biddulph

File:8 bytes vs. 8Gbytes.jpg *Source*: http://en.wikipedia.org/w/index.php?title=File:8_bytes_vs._8Gbytes.jpg *License*: Creative Commons Attribution 2.0 *Contributors*: Daniel Sancho from Málaga, Spain

File:Flash memory cards size.jpg *Source*: http://en.wikipedia.org/w/index.php?title=File:Flash_memory_cards_size.jpg *License*: unknown *Contributors*: Afrank99, Ivob, MMuzammils, Moxfyre, Rhe br, Solomon203, Warden, Zxb, 1 anonymous edits

File:SD-microSD adaptor.jpg *Source*: http://en.wikipedia.org/w/index.php?title=File:SD-microSD_adaptor.jpg *License*: Creative Commons Attribution-Sharealike 3.0,2.5,2.0,1.0 *Contributors*: Yoskov

File:SD card pinning.jpg *Source*: http://en.wikipedia.org/w/index.php?title=File:SD_card_pinning.jpg *License*: Creative Commons Attribution-Sharealike 3.0 *Contributors*: original photo: User:Afrank99, derivative work: User:Jaho

File:Mini SD card pinning.jpg *Source*: http://en.wikipedia.org/w/index.php?title=File:Mini_SD_card_pinning.jpg *License*: Creative Commons Attribution-Sharealike 3.0 *Contributors*: User:JohnHind

File:Micro SD card pinning.jpg *Source*: http://en.wikipedia.org/w/index.php?title=File:Micro_SD_card_pinning.jpg *License*: Public Domain *Contributors*: JohnHind

File:Sd insides.png *Source*: http://en.wikipedia.org/w/index.php?title=File:Sd_insides.png *License*: Creative Commons Attribution-Sharealike 3.0 *Contributors*: User:Saltmiser

File:SD-extreMEmory 2GB alt innen.jpg *Source*: http://en.wikipedia.org/w/index.php?title=File:SD-extreMEmory_2GB_alt_innen.jpg *License*: Creative Commons Attribution-Sharealike 3.0 *Contributors*: Manorainjan

File:SDHC inside.jpg *Source*: http://en.wikipedia.org/w/index.php?title=File:SDHC_inside.jpg *License*: Creative Commons Attribution-ShareAlike 1.0 Generic *Contributors*: George Shuklin

File:Sd4gy crop.jpg *Source*: http://en.wikipedia.org/w/index.php?title=File:Sd4gy_crop.jpg *License*: Creative Commons Attribution-Sharealike 3.0 *Contributors*: Wirepath

CPSIA information can be obtained at www.ICGtesting.com
Printed in the USA
LVOW041836270912

300607LV00009B/158/P